BACKYARD BALLISTICS

BACKYARD BALLISTICS

Build potato **cannons,** paper match **rockets,** Cincinnati **fire kites,** tennis ball **mortars,** and more **dynamite** devices

William Gurstelle

CHICAGO
REVIEW
PRESS

Library of Congress Cataloging-in-Publication Data

Gurstelle, William.
    Backyard ballistics : build potato cannons, paper match rockets,
Cincinnati fire kites, tennis ball mortars, and more dynamic devices /
William Gurstelle.
        p. cm.
    Includes index.
    ISBN 1-55652-375-0
    1. Science—Experiments. 2. Ballistics—Experiments. I. Title.
Q164 .G88 2001
531'.55'078—dc21

                                    2001017321

Visit the *Backyard Ballistics* Web site for additional help or questions
at www.backyard-ballistics.com.

Designed by Lindgren/Fuller Design

©2001 by William Gurstelle
All rights reserved
First edition
Published by Chicago Review Press, Incorporated
814 North Franklin Street
Chicago, Illinois 60610
ISBN 1-55652-375-0
Printed in the United States of America

This book is dedicated to my father,
Harold H. Gurstelle.

At other times, but especially when my uncle Toby was so unfortunate as to say a syllable about cannons, bombs, or petards—my father would exhaust all the stores of his eloquence (which indeed were very great) in a panegyric upon the Battering-Rams of the ancients—the Vinea which Alexander made use of at the siege of Troy.

He would tell my uncle Toby of the Catapultae of the Syrians, which threw such monstrous stones so many hundred feet, and shook the strongest bulwarks from their very foundation—he would go on and describe the wonderful mechanism of the Ballista which Marcellinus makes so much rout about; the terrible effects of the Pyraboli, which cast fire; the danger of the Terebra and Scorpio, which cast javelins...

But what are these, would he say, to the destructive machinery of Corporal Trim? Believe me, brother Toby, no bridge, or bastion, or sally-port, that ever was constructed in this world, can hold out against such artillery.

—Laurence Sterne,
The Life and Opinions of
Tristram Shandy, Gentleman

Contents

Acknowledgments

Thanks to Hank Childers, Frank Clancy, Peter Gray, Randy Tatum, Tom Tavolier, Clark Tate, the helpful friends of my sons who took special interest in creating this work, as well as many friends and relatives for their assistance and encouragement. Todd Keithley at Jane Dystel Literary Management deserves special recognition for making *Backyard Ballistics* a reality. Special thanks to my wife Barb, and my sons Andy and Ben for their help and inspiration.

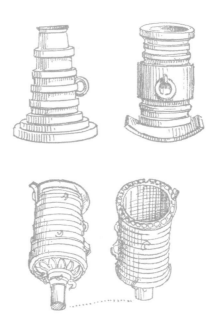

Introduction

THE BOY MECHANIC

I have a special shelf in my library with about half a dozen books of science projects. In these books, from the 1892 *All-American Boy's Handy Book* through the 1939 *Fun for Boys*, there are hundreds of complicated projects that modern kids could hardly fathom. Most kids today will not be able to turn parts on a lathe or practice their taxidermy skills on a raccoon. But I love these books for the ideas they contain: how to build wooden street racers, how to build "superhet" radios, and how to construct and wind electric motors.

One of the best books I own is the 1913 edition of *The Boy Mechanic: 700 Things for Boys to Do. Popular Mechanics* magazine published it when electricity and aviation were young, when most people didn't even own cars. The cover shows a boy stepping off a cliff in a glider (built according to instructions on page 171). The image would make publishers and parents nervous today, but there were far fewer lawsuits in 1913. Back then it was your own carelessness, not *Popular Mechanics*', if you didn't make your glider's joints fast and strong.

Paging through the book, several things become apparent. Electrical projects are prominent. You can build an "electric bed warmer," an electric toaster, or a wireless telegraph. Magic and illusions are also featured—you can build an "electric illusion box" with a secret compartment that magically transforms a full red apple into a half-eaten one before your eyes, or you can learn a whole bunch of coin tricks.

The Boy Mechanic is also long on things that shoot or explode. For instance, the gunpowder-driven "Fourth-of-July Catapult" flings a full-sized mannequin a hundred feet into the sky! The excitement goes on for 460 pages: how to build a working cannon, how to fashion a crossbow, and how to set a smoke screen.

The book contains safer but less exciting ideas, too—for example, how to make a lamp or a belt hanger. But that's not why boys and girls used to read such books. They read them to channel their excitement and take risks in a productive way. They read them because, for yesterday's young people, this sort of experimentation was a normal and expected part of growing up. *Backyard Ballistics* is the direct descendant of those books. Obviously, the time for making needle-tipped blowguns or mixing a homemade batch of mercury fulminate has passed. *Backyard Ballistics* proudly wears the mantle of those books and fits the needs of the present time. No cerebral computer simulations here, just plenty of real fun with a good purpose.

BE AN AMATEUR, PROUDLY

Around 1700, the Italian composer and music teacher Arcangelo Corelli published his sonatas for violin. The sonatas were bound in a stunningly beautiful book, printed on the finest quality paper of the day. The music was beautiful, too, despite its simplicity; the sonatas contained simple melodies without ornamentation or embellishment.

This simple music was fundamentally important to a generation of violinists—perhaps the finest generation ever to work a bow. You see, Corelli made the music simple for a reason: he expected his violin students to be amateur composers as well. The students themselves completed the music in his sonatas; they were supposed to improve his lines with trills and musical runs of their own. They interpreted the dynamics and phrasing of individual passages, added their counterpoint harmonies, and so on.

This was the age of the amateur, an era that produced tremendous creativity and genius. An amateur is literally someone who loves what he or she does. An amateur does something because he or she wants to, not because he or she is paid to. Professionalism, in the modern sense of the word, was almost unknown. And that's what drove eighteenth-century genius. Whether you played music or created inventions, if you were great, your work evidenced your passion.

A different way of thinking changed the nineteenth century. The improvisation and individualism so important to virtuosos of the day were eclipsed by rigid adherence to rules and uniformity. A new breed of professional musician came to the forefront in the late nineteenth century. The composer carefully detailed all the dynamics, accents, and tempos for musicians as musical improvisation and interpretation took a back seat to technique and cold precision.

Like music, inventing and engineering metamorphosed into professions of intellectualism, methodology, and procedure during the 1830s and '40s. Of course, they had to. Prior to the 1800s, every tool and fixture was handwrought by craftsmen using techniques honed by years of apprenticeship and experience. The Industrial Revolution changed this. Mass-produced goods created in the steam-driven factories and mills of the industrial Northeast required the coordinated efforts of many dedicated professional engineers—of experts who would do things precisely and methodically every time.

Today, there are times and places for both professionals and amateurs. As we enter the twenty-first century, the influence of the amateur has eroded, but the amateur spirit lives on.

For example, NASA put the space shuttle into the sky, but who made air transportation a reality? Orville and Wilbur Wright's invention of the airplane was essentially a two-man effort. In December of 1903, at a remote, windy beach in North Carolina, the airplane was born when first Orville and then Wilbur took off and landed in a flying machine. All of the major components of this machine—wings, propeller, gasoline engine—were of the Wrights' own design.

The Wrights were self-taught aeronautical engineers of the first order. They calculated that their flying machine's engine would need to develop eight horsepower at the propeller in order to attain enough speed to take off, but the engine could weigh no more than 200 pounds. They wrote to automobile companies, hoping to purchase an engine that could meet their requirements. No engine could, so Orville and Wilbur, with the help of a single machinist, designed, built, and tested an engine that could do the job. These three amateurs went from concept to completion in just six weeks.

The first picture of the Wright brothers' airplane in flight was taken by a local Kitty Hawk resident named John Daniels. When the contraption actually left the ground, Daniels got so excited that he forgot his task of taking the photograph. He didn't think he'd squeezed the bulb. But fortunately, he did.

In the moment captured in this first picture, you can see Wilbur running. He's caught in mid-stride. Orville is on the machine, and the machine is off the ground. The flight's distance was 120 feet. Later that same day, they made a flight of over 800 feet.

It took years of experimentation, research, and building for the Wrights to build a flying machine capable of powered heavier-than-air flight. But fly they finally did, and these two amateurs ushered in the age of powered flight.

This book is about being creative in the name of science and experimentation. The experiments are fun, and that's reason enough to try them. But, there's a strong possibility that you might learn something while you're at it!

The creative part of being an engineer happens only in those moments when engineers lay their practiced, detached professionalism aside and behave like amateurs. Invention, by its nature, lies outside the professional's mind-set of established knowledge and moves into limited and reasoned risk taking. In *Backyard Ballistics*, these experiments will open your eyes to new concepts, and then perhaps to other ideas not explicitly explained in the book. If you come up with a good one, I hope you'll write to me.

Build and learn. This book is for you, the amateur scientist.

Time Line

212 B.C. Siege of Syracuse by Marcellus. Death of Archimedes.

190 B.C. Hipparchus develops the foundations of trigonometry.

50 B.C. Roman catapults are in widespread use.

A.D. 550 Byzantines develop Greek fire and other incendiaries.

800 Chinese use primitive shooting mortars against invaders.

1250 English monk Roger Bacon popularizes gunpowder in Europe.

1300 Crossbows are in widespread use.

1304 The largest of all siege weapons, Ludgar the Warwolf, is built.

1642 Isaac Newton is born in England.

1687 *Philosophiae Naturalis Principia Mathematica* is published by Isaac Newton.

1783 First lighter-than-air flying craft is invented by the Montgolfiers.

1846 Nitroglycerin is invented.

1867 Dynamite is invented.

1892 Calcium carbide is first manufactured.

1903 Wright brothers develop the airplane.

1944 V-2 rocket is launched from occupied Europe.

1955 Los Alamos Laboratories sends first man-made object into space.

1957 Sputnik is launched into orbit by Russian Space Agency.

BACKYARD BALLISTICS

▷ 1 ◁

Keeping Safety in Mind

GENERAL SAFETY RULES

When you were a child, people told you not to play with matches for a good reason—they can be dangerous! If you don't follow the directions closely, any of the experiments in *Backyard Ballistics* could cause harm to you and your possessions. Remember to always follow the instructions closely. Do not make changes to the materials or construction techniques. It can lead to unexpected and unintended results.

A Very Important Message
The projects described in the following pages have been designed with safety foremost in mind. However, as you try them, there is still a possibility that something unexpected may occur. It is important that you understand neither the author, the publisher, nor the bookseller can or will guarantee

your safety. When you try the projects described here, *you do so at your own risk.*

Some of these projects have been popular for many years, while others are new. Unfortunately, in rare instances, damage to both property and people occurred when something went wrong. The likelihood of such an occurrence is remote, as long as the directions are followed, but remember this—things can go wrong. Always use common sense and remember that all experiments and projects are carried out at your own risk.

Be aware that each city, town, or municipality has its own rules and regulations, some of which may apply to the projects described in *Backyard Ballistics*. Further, local authorities have wide latitude to interpret the law. Therefore, you should take the time to learn the rules, regulations, and laws of the area in which you plan to carry out these projects. A check with local law enforcement will tell you whether the project is suitable for your area. If not, there are other places where experiments can be undertaken safely and legally. If in doubt, be sure to check first!

Ground Rules

These are your general safety rules. Each chapter also provides specific safety instructions.

1. The experiments described here run the gamut from simple to complex. All are designed for adults or, at a minimum, to be supervised by adults. Take note: Some experiments involve the use of matches, volatile materials, and projectiles. Adult supervision is mandatory for all such experiments.

2. Read the entire project description carefully before beginning the experiment. Make sure you understand what the experiment is about, and what it is that you are

trying to accomplish. If something is unclear, reread the directions until you fully comprehend the entire experiment.

3. Don't make substitutions for the specific liquids and aerosols indicated for use in each experiment. Stay away—far away—from gasoline, starting ether, alcohol, and other powerful inflammables. There are few things as dangerous as flammable liquids or aerosols. They can and do explode, and the consequences can be disastrous.

4. Use only the quantities of fluid listed in the project instructions. Don't use more propellant than specified.

5. Don't make substitutions in materials or alterations in construction techniques. If the directions say to cure a joint overnight, then cure it overnight. Don't take shortcuts.

6. Read and obey all label directions when they call for materials such as PVC cement, primer, and other chemicals.

7. Remove and safely store all cans or bottles containing naphtha, hairspray, or any other flammable substance prior to performing the experiment. A good rule of thumb is to maintain a hazard-free radius of at least 50 feet around the area in which you plan to work.

8. The area in which the projects are undertaken must be cleared of all items that can be damaged by projectiles, flying objects, and so forth.

9. Keep people away from the firing zone in front of all rockets, mortars, cannons, etc. Use care when transporting, aiming, and firing, and always be aware of where the device is pointing.

10. Wear protective eyewear when indicated in the directions. Similarly, some experiments call for hearing protection, blast shields, gloves, and so forth. Always use them.

Remember this:

▶ The instructions and information are provided here for your use without any guarantee of safety. Each project has been extensively tested in a variety of conditions. But variations, mistakes, and unforeseen circumstances can and do occur. Therefore, all projects and experiments are performed at your own risk! If you don't take this seriously, then put this book down; it is not for you.

▶ Finally, there is no substitute for your own common sense. If something doesn't seem right, stop and review what you're doing. You must take responsibility for your personal safety and the safety of others around you.

WORKING WITH PVC PIPE

Several of the projects contained here involve cutting and joining PVC pipe. This section tells you what you need to know in order to make safe and secure joints.

First, you should be aware that there are at least four types of plastic pipe and plastic pipe joints available: PVC, CPVC, ABS, and PB. The letters are abbreviations for the type of plastic material that composes the pipe. Pressure rated, schedule-40 PVC pipe and pipe fittings are made of white polyvinyl chloride. This is the type of pipe and pipe joints recommended for these projects.

Cutting and Fitting PVC Pipe

PVC pipe is easily cut with a regular, fine-bladed handsaw. It is important that all the cuts be made as close to 90 degrees to the centerline of the pipe as possible. That way, you won't leave any interior gaps, which will weaken the joint.

You may want to "dry fit" the pipe into the joints before you apply any cement to see how things fit. Sometimes the dry-fitted pipes and joint fittings stick together so tightly it is hard to get them apart. If that happens, you can carefully whack the fitting loose with a wooden block.

Joining and Cementing PVC Pipe

The process of joining and cementing PVC pipe is technically called "solvent welding." The solvent melts the plastic so when you push the pipe and the pipe fitting together, the two parts fuse as the solvent evaporates. Each type of plastic pipe has its own special solvent. Some solvents are advertised to work on several types of plastic, but it is strongly recommended that you use the solvent that is meant solely for the type of plastic you're working with. At the hardware store the solvent you need is called "PVC cement."

The solvent works only on *clean* surfaces—surfaces with no dirt, no grease, and no moisture. Wipe the inside of the fitting and the outside of the pipe with a clean cloth. Then, apply PVC primer (called "purple primer") to the ends.

Next, coat the surfaces that you want to join with a liberal amount of PVC cement. (PVC cement, which is a solvent, should only be used in well-ventilated areas.) Push the pipe into the pipe fitting quickly and give it a one-quarter turn as you seat it. Hold it tight for about 15 seconds, and then, voilà!—you are done. Be sure to observe the cure times shown on the PVC cement can's directions.

▷ **2** ◁

The Potato Cannon

The potato cannon, or spud gun as it is sometimes called, is nearly legendary in amateur science circles. You'll be amazed at how easy it is to make a working ballistic device out of materials readily available at your local hardware store. Don't worry, the potato cannon doesn't use dangerous gunpowder or rocket fuel to blast the potato in the air. Instead, the cannon takes advantage of the considerable energy contained within the aerosol propellant of common hairspray.

Thousands of people, from adolescent boys and girls to serious experimenters at Ivy League universities, enjoy shooting homemade ballistic devices like this. It's appealing for several reasons. First, the cannon is simple to build. Second, it really does work well. And finally, it's easy to understand. Unlike the complicated configuration of a computer's silicon chips, the average person can figure out (with the help of this book) the physics of the cannon.

2.1 Completed spud gun

People love making the potato cannon. Don't be too surprised if the hardware store clerk takes a quick look at your materials and says, "Making a spud gun, eh?" It happens to me all the time.

Building the Potato Cannon

Working with PVC Pipe
PVC pipe is the greatest home plumbing invention of the twentieth century. Unlike heavy steel pipe, the average person can quickly cut, join, and fasten PVC pipe with a minimum of materials and a small amount of practice. This makes it the perfect spud gun raw material.

THE PIPE

PVC pipe is made of a polyvinyl chloride plastic. Manufacturers make these pipes in various thicknesses. You specify the thickness by referring to its "schedule." For our experiment, we need schedule-40 PVC pipe. It also comes in a variety of diameters: 1-inch, 2-inch, and so on. Buy it in 8-foot lengths and cut it to the size you need with a hacksaw.

THE CONNECTORS

PVC pipe manufacturers make a variety of connectors to join pipes in the way plumbers (and spud gunners) need. Couplings join pipes of similar sizes. Threaded couplings have female pipe threads cut into at least one end. Reducing bushings join a pipe of one size to a pipe of a smaller size. End caps simply cap the end of the pipe.

JOINING THE PVC

The insides of the connectors are either smooth or cut with screw threads. Sometimes we'll want to join two smooth pieces, which can be "solvent welded" together using special PVC cement. (Note: Always use special-purpose PVC cement on PVC pipes and connectors. Regular glue won't work.) Other times, we'll want to join two threaded pieces that can simply be screwed together.

Go to the local hardware store's plumbing section and ask the clerk to help you find the items on page 12. Yes, the big commercial hardware stores usually have all of these items (except the lantern sparker and hairspray). However, I recommend going to your local hardware store because the clerks are usually much more helpful. Sometimes, they will even cut the pipe to size for you and not charge you for a full 8-foot piece of pipe.

EXPLOSIVES

Explosives are chemicals used mostly in commercial and military applications to induce the production of hot, rapidly expanding gas. The rapidly accelerating gas is powerful enough to reduce rock to rubble or flatten buildings for demolition.

Explosives are not the same as propellants. Gunpowder and Pyrodex (used in rifles and shotguns) are examples of propellants. These substances burn vigorously, but in a slower and more controlled fashion than explosives.

There are many different types of explosives. "High explosives" are used mostly in military applications. High explosives include trinitrotoluene (TNT) and cyclotrimethylenetrinitamine (RDX). Plastic explosives are pliable compounds that can be molded into desired shapes for special-purpose applications.

Commercial explosives are less powerful than high explosives. The first commercial explosive was nitroglycerin. Later, commercial blasting operations began to use safer and more controllable substances such as ANFO (see below) and dynamite.

In mining operations in the western United States, it is often necessary to remove huge quantities of rock and soil that cover seams of coal. The common method of removing the "overburden" of rock is to drill a long series of holes at close intervals in the rock to be removed. The holes are filled with a mixture of ammonium nitrate and fuel oil. This mixture is called an ANFO explosive. A fuse made of a special type of inflammable cord and a primer charge is placed in the ANFO mixture and ignited. Large mining operations produce blasts large enough to be measured by seismographs around the world. In fact, mining companies must alert certain foreign governments prior to these huge mountain-moving blasts, or else other countries might think illegal nuclear explosives testing is taking place.

All explosives, fireworks included, are very dangerous in the hands of the untrained. Experienced miners and military experts know and understand the detailed information required to use them safely and effectively; the rest of us will have to confine our experience with explosives to reading about them.

Materials

- Hacksaw

- Shaping file

- (1) 36-inch length of 2-inch diameter schedule-40 PVC pipe

- (1) 3- to 2-inch diameter reducing bushing

- (1) 14-inch length of 3-inch diameter schedule-40 PVC pipe

- (1) can PVC primer

- (1) can PVC cement

- (1) 3-inch coupling, one side smooth, one side threaded

- Electric drill with $\frac{1}{8}$-inch drill bit, $\frac{5}{16}$-inch drill bit

- (1) flint and steel lantern sparker. (This small device is widely available at most camping goods stores and large discount stores with camping equipment departments. It is designed to ignite the mantles of lanterns. It consists of a steel wheel that is rotated against a flint by means of a knurled brass handle. It generally retails for less than five dollars.)

- Large adjustable wrench

- Duct tape

- (1) 3-inch diameter threaded PVC end cap

- (1) 4-foot length of 1-inch diameter wooden dowel or broom handle

- Hairspray in a large aerosol can (Be sure it's an aerosol can and not a pump spray. Spud gunners typically buy the most inexpensive brand of hairspray. Our interest is in its hydrocarbon propellant, not its holding power or scent.)

- Protective gear including safety glasses, earplugs, and gloves

- Bag of potatoes

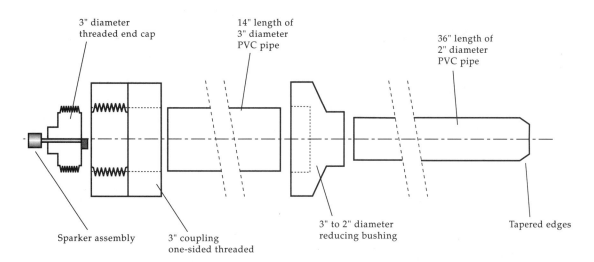

3" diameter
threaded end cap

14" length of
3" diameter
PVC pipe

36" length of
2" diameter
PVC pipe

3" to 2" diameter
reducing bushing

Tapered edges

Sparker assembly

3" coupling
one-sided threaded

2.2 Assembly drawing

Place all of your materials and tools in front of you. Crafting a spud gun from raw materials takes at least two hours of filing, cutting, and drilling. You may need an extra pair of hands to hold things in place while you are gluing. After the pieces are put together, you'll need to let the PVC cement cure overnight.

1. Use the hacksaw to cut the PVC pipes to the desired lengths.
2. Use the file to taper one end of the long, 2-inch diameter pipe section so it forms a sharp edge. The edge will cut the potato as it is rammed into the muzzle of the gun.
3. Use PVC primer before cementing. Attach the 3-inch side of the 3- to 2-inch reducing bushing to one end of the 3-inch pipe using the PVC cement. Be sure the joints are clean and that you apply the cement according to the directions on the can. Don't forget to observe the

2.3 Applying the PVC cement

directions for curing times. You MUST let all the connections cure overnight in a well-ventilated area.

4. Carefully cement the smooth, unthreaded side of the 3-inch,
   one-sided threaded coupling to the 3-inch PVC pipe. Do
   not get any cement on the exposed pipe threads. If you
   do, you won't be able to screw the end cap into place.

5. The 36-inch long, 2-inch diameter pipe is the muzzle of
   the potato gun. Cement the untapered side to the 2-inch
   side of the reducing bushing.

6. Carefully drill a hole large enough for the sparker (usually ¼ inch or ⁵/₁₆ inch, but match the twist drill you use
   to the diameter of the sparker's hollow bolt) to snugly fit
   through the middle of the 3-inch threaded end cap.

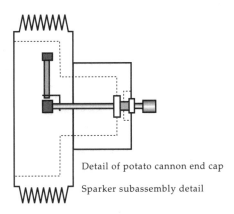

Detail of potato cannon end cap

Sparker subassembly detail

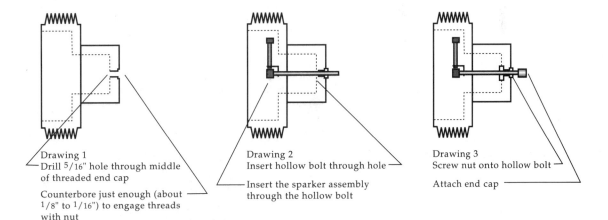

Drawing 1
Drill 5/16" hole through middle
of threaded end cap

Counterbore just enough (about
1/8" to 1/16") to engage threads
with nut

Drawing 2
Insert hollow bolt through hole

Insert the sparker assembly
through the hollow bolt

Drawing 3
Screw nut onto hollow bolt

Attach end cap

2.4 Detail for mounting sparker inside the end cap

7. Take the hollow, threaded bolt assembly from the sparker and insert it through the hole made in step 6. Depending on the type of sparker you have, you may have to drill (counterbore) a $^3/_8$-inch diameter depression on the outside surface of the end cap. Make it $^1/_{16}$- to $^1/_8$-inch deep in the PVC. You just need to reach and engage the hollow bolt's screw threads with the nut. (See diagram 2.5.)

8. Mount the sparker by unscrewing the knurled end cap from the shaft. Be aware there is a spare flint inside the end cap, so watch for it. Unscrew the nut and remove the metal

angle piece. (Throw away the angle piece.) Insert the sparker shaft through the hole and tighten the nut until the sparker is firmly in place. The shaft will slide in and out, but it won't come out. Replace the end cap and tighten the lock screw.

9. Allow the entire assembly to cure overnight. Do not test the sparker until the assembly is fully cured.

10. For extra safety, wrap the barrel and joints with multiple layers of duct tape (excluding the threaded end cap).

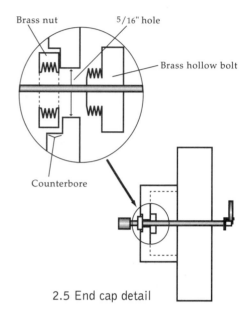

Brass nut   5/16" hole

Brass hollow bolt

Counterbore

2.5 End cap detail

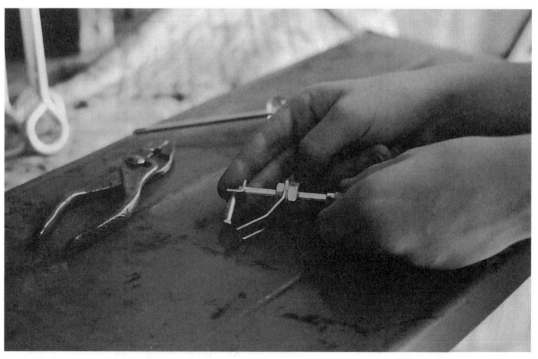

2.6 Sparker

ALFRED NOBEL

Back in the old days, mining was tough work. In order to obtain ore, men had to swing picks and hammers to crack the rock so they could haul it for refining and smelting. When gunpowder came along, the miners thought that maybe it could be used to break up the rocks. But gunpowder wasn't powerful enough to do the job. Then an Italian scientist, Ascanio Sobrero, invented nitroglycerin, a very powerful liquid explosive.

Nitroglycerin is extremely sensitive, making it difficult to transport and use safely. A slight jolt or even a spark from friction can cause it to explode.

The Swedish scientist Alfred Nobel developed a way to mix the unstable nitroglycerin with clay and other binder compounds, creating the first explosive that was safe to transport and, with proper care, safe to use. The mixture he devised was called dynamite.

Nobel was farsighted enough to realize that his invention would not only revolutionize commercial mining but warfare as well. The use of his invention in war troubled him greatly. As a young man, he had thought that his invention would make wars less likely because the new possibility of immense destruction would dissuade countries from going to battle. But as he grew older, he realized his invention only made war easier. For this reason, Nobel wanted all the money he made from dynamite to be used in the interest of peace and scientific advancement. The Nobel Prizes were organized and funded by Nobel's estate to recognize those people who make the greatest strides toward peace and science, toward the end of war.

KEEPING SAFETY IN MIND

1. The potato cannon shoots with enough force to cause injury. Always use extreme care when aiming the device. Make certain the end cap is fully screwed on.
2. The firing will cause a small recoil. You will need to mount the cannon securely to its firing platform.
3. Check the cannon after every use for signs of wear and to make sure the barrel maintains its structural integrity. Replace any worn sections or parts immediately.
4. Use only the type and quantity of propellant described. Do not use too much propellant or you may damage the cannon. Make certain the hairspray can is removed to a safe distance before the cannon is fired.
5. This device produces a loud noise. Use protective eyewear, hearing protection, and protective gloves.
6. Clear the area in front of the cannon for 200 yards. Clear the area behind the cannon for at least 25 yards.
7. Yell "Spuds away!" or "Fire in the hole!" before shooting to make sure no one's about to walk into the field of fire.

FIRING THE POTATO CANNON

It's finally time. Your cannon is ready, and you've studied the safety procedures. You've got a 10-pound bag of fresh Idaho russets, an economy-sized can of hairspray, and an itchy trigger finger. Let's march on out to the testing field and send those tubers into the stratosphere.

1. Remove the end cap.
2. Using the dowel or broom handle, ram a potato into the cannon from the muzzle end. The cutting edge made in step 1 of the assembly will cut the potato into a plug of

the correct size. The potato must fit snugly on all sides of the muzzle. Any gaps will allow the expanding gas to "blow by" the potato. If that happens, the potato plug won't go far.

3. Use the dowel to push the potato projectile 30 to 32 inches down into the cannon muzzle.

4. Direct a stream of hairspray into the firing chamber of the cannon (the 3-inch diameter cylinder where the sparker is mounted). It is important that you introduce the correct amount of hairspray into the combustion chamber. Because the amount of spray delivered per unit varies between spray cans, the correct amount of hairspray propellant is determined by trial and error. Start by using a very short burst of hairspray and increase the amount by small intervals until maximum performance is attained. *To reiterate*: Start with a very short burst and then try progressively larger amounts of hairspray. In general, a burst length of about two econds works well. *However*, the amount of hairspray actually delivered will vary among cans of hairspray. Therefore, start small and work up, but don't exceed two seconds.

5. Immediately replace the end cap and screw it on securely.

6. Twist the igniter sharply to fire the cannon.

TIPS AND TROUBLESHOOTING

1. A support, such as a ladder mount shown on page 20, is required to securely hold the cannon.

2. Use fresh potatoes. Old potatoes tend to mush or shred when rammed into the tapered muzzle. This results in the "blow by" effect described earlier.

3. Clean the spud gun after every few shots. Use the dowel

2.7 Using a ladder mount to secure the potato cannon

to push a wet rag through the muzzle to remove potato
and hairspray residue.

4. **Note:** The threaded end cap often becomes difficult
to remove after firing. Keep the threads scrupulously
clean and always have a wrench handy in case it
becomes stuck.

WHAT'S GOING ON HERE

You may be wondering just what is happening inside the cannon. Twisting the knob causes sparking inside the cannon's
firing chamber. The sparks ignite the hairspray–air mixture
inside. The gaseous mixture expands quickly and pushes
against everything as it expands. The rigid PVC walls won't
move, but the potato can, and does. In fact, the expanding
gas moves so quickly and so rapidly that the potato flies out
of the tube and into the air.

One can almost imagine Sir Isaac Newton, in a velvet waist-coat, breeches, and powdered wig, flicking the firing knob of a potato cannon. "I say," Newton would remark, "this proves my theories better than ye falling apples." Newton, of course, was a seventeenth-century English physicist and mathematician. He was also the fellow who came up with the concept of gravity while supposedly resting under an apple tree. In addition to his work on gravity and astronomy, Newton published a seminal book in terms of understanding the nature of motion. He titled it *Philosophiae Naturalis Principia Mathematica*. In *Principia*, he described three immutable physical laws that explain what are called the statics and dynamics of objects. Statics is the study of forces on an object at rest; dynamics is the study of how forces affect that object in motion.

You've probably heard of Isaac Newton's three laws of motion. These laws are the foundation of dynamics. The first law says that once something starts to move, it stays in motion until some force acts to stop it. This is often described as the principle of inertia. The second law says that if a force is exerted on an object, it goes faster, and the amount of force applied is proportional to the amount it accelerates. The final law states that every action has an equal and opposite reaction.

Our potato cannon illustrates parts of all three laws. The first law of motion tells us that after the potato is launched, it will continue to shoot forward until another force stops it. If the spud hits a tree, the opposing force is obvious. But if there is no tree in its way, why does the potato stop at all? Why doesn't the potato keep shooting forward forever? The potato stops moving only when the frictional force of the air and ground slow the potato and eventually stop it.

Let's go back to the example of a spud hitting a tree. We might wonder how much force the potato experiences when it thunks against the trunk. Well, according to Newton's second law, that thunk—the force of the potato—is equal to the mass of the potato multiplied by the acceleration of the potato. The spud gun imparts a big acceleration during the time the potato moves through the barrel of the cannon. So, when the potato decelerates from blasting through the air at high speed to an abrupt stop at the tree trunk, that's one big force encountered by the tree. What all this means is that the mass of the potato times its acceleration equals a huge force when it hits the tree. Bang!

Here is a question for science enthusiasts: If we make the cannon barrel longer, will the potato go farther? It seems logical that it would. If the barrel is longer, the ignited and expanding hairspray gas would push on the potato for a longer time. This should impart a greater force and make the potato go farther.

Sounds reasonable, right? So, what's the perfect barrel length? Well, that's for you to find out. A few experiments with barrels of varying lengths should provide the answer. Check out "Ideas for Further Study" at the back of this book for information on how an engineer would answer this question using the Scientific Method.

The third law says that when the potato is fired from the cannon muzzle, an equal and opposite reaction will be exerted on the support structure holding the cannon. This phenomenon is called "recoil" or "kickback." This is why it is important to secure the cannon to the launch platform.

▷ 3 ◁

Back Porch Rocketry

This chapter is all about chemical-free rocketry. Unlike model rocketry (which, incidentally, is a lot of fun), these projects don't use manufactured chemical rocket engines. Instead, we'll cut and paste paper and cardboard together and figure out how many different ways we can send our constructions airborne.

Let's head to the back porch and make something fly.

THE PAPER MATCH ROCKET

Got a minute? In the mood for a bit of pyrotechnic handwork? Try building a paper match rocket. It's an uncomplicated device that can be built in a matter of minutes with the simplest of materials.

The match rocket illustrates the principle of rocket propulsion. In significant ways, this tiny device is similar to the

3.1 Paper match rocket

space shuttle's booster rockets. If you don't believe that a match, a piece of aluminum foil, and a couple of pins can blast off, read on.

Building the Paper Match Rocket

You create the rocket by attaching a ported metal foil over a match head. Ported is an engineer's way of saying that the foil is wrapped around the match head so that only two small holes in the foil remain. The holes are like the exhaust nozzles of a NASA rocket. When everything's in place, you'll light the match head. This in turn will cause a tiny pocket of high-pressure gas to form inside the foil. The gas will shoot out of the aluminum exhaust ports. Zip! We have liftoff!

Materials

Rocket

- (1) book of paper matches
- Scissors
- Aluminum foil
- (2) pins or sewing needles

Launcher

- (1) 1-inch length of ¼-inch diameter copper tube
- "T" hinge, about 2 inches long x 1½ inches wide
- Adhesive or brazing
- Wooden block, at least 3 inches x 3 inches and ³⁄₄ inch thick (large enough to accommodate the hinge and thick enough to accommodate the wood screw)
- (1) ½-inch-long wood screw
- Paper clip (optional)
- Cigarette lighter or wooden match
- Safety glasses

Making the Rocket

1. Remove a single match from the matchbook. Use scissors to trim off the torn end squarely.
2. Cut a piece of aluminum foil slightly less than 1 inch square.
3. Place the foil square on a flat surface. Next, place the pins on either side of the match head, as shown in diagram 3.2. The pins act as a form or mold for the holes that make the jet exhaust ports.

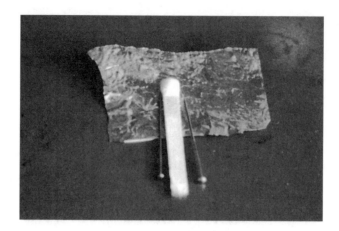

3.2 First, place pins and match on foil square.

4. Wrap the foil around the head of the match, extending the foil between ¼ and ½ inch past the head. Take your time and use care as you mold the foil around the match and pins. The jet ports must be perfectly formed in order for the rocket to work.

5. Carefully remove the pins from the rocket assembly. The cylindrical areas left after removing the pins form the jet exhaust ports.

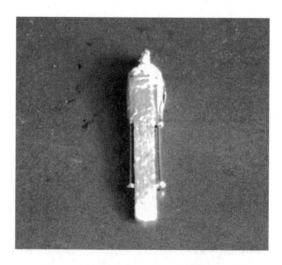

3.3 Wrap foil tightly around the pins and match.

3.4 Carefully remove pins, leaving jet ports open.

Making the Launcher

1. Make a platform for the launcher. Attach the ¼-inch copper tube to a hinge by using adhesive or brazing. Attach the hinge/tube assembly to the small wooden block to use as a launching platform. To improvise a mount that can be aimed, insert a wood screw halfway into the platform under the swinging end of the hinge as shown in diagram 3.5. Adjust the height of the screw to launch the rocket higher or lower.

2. If building this launcher seems like too much work, you can make a fairly serviceable launcher by simply bending a paper clip into a launcher, although the performance of the rocket won't be the same. There are many ways to bend a paper clip into a serviceable launcher; experiment to see what works best.

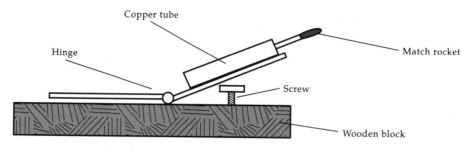

Copper tube

Hinge

Match rocket

Screw

Wooden block

3.5 Match rocket launcher assembly

Launching the Rocket

Place the rocket in the firing tube. Hold the match rocket head in the flame of a cigarette lighter or a wooden match. After a few seconds the match head will ignite. The burning match head will force out hot, expanding gases from the jet exhaust ports at high velocity and propel the rocket forward.

KEEPING SAFETY IN MIND

1. The match rocket head becomes quite hot after ignition. The rocket head is hot enough to melt carpeting, burn a hole in your good shirt, and maybe even cause a minor burn.

2. After ignition, the rocket launches into uncontrolled flight. Uncontrolled flight plus hot match heads equals an imperative to wear safety glasses.

3. Match rockets can fly for over thirty feet. Therefore, remove any hazardous material in a 30-foot radius from the launch pad.

4. For all of the reasons stated above, you must perform this experiment outdoors.

TIPS AND TROUBLESHOOTING

1. Making match rockets is an art as well as a science. It takes a keen eye and a steady hand to make a match rocket that flies well.

2. You may encounter the blowout problem. Blowout occurs when the igniting match blows a hole through the aluminum foil cover. If that happens, the gas inside the foil won't get channeled through the jet exhaust ports. A blown-out rocket will not fly.

3. You're also likely to encounter a jet port blockage problem. The narrow jet exhaust ports are delicate and easily crimped if you're not extremely careful when removing the pins from the foil head. Sometimes you'll remove the pins with no problem, and then crimp the jet ports simply by handling the rocket too roughly with your hands.

4. How far will the match rocket fly? Some people claim their best match rockets have flown over 50 feet. The distance will depend on how well you make the rocket, the weight of the match rocket, the amount of chemical energy stored in the match head, and so on. These variables open up the doors for further experimentation. For instance, can you shave the matchstick to make it lighter and still get a fairly straight flight? The matchstick is

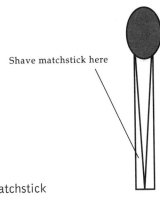

Shave matchstick here

3-6 Matchstick

important because it provides the rocket with a stable center of gravity. (If the match head did not have a stable center of gravity, it would just spin and tumble after ignition and not really go anywhere.) But the matchstick is heavy relative to the match head. By shaving the matchstick into a narrower shape, as shown in diagram 3.6, the match rocket will fly far and true.

5. There are many other variables to explore. Experiment with different types and colors of match heads to see if some have more power than others. Find the angle of launch (trajectory) that maximizes the distance traversed.

Real rocket performance analysis looks at a host of highly technical variables such as the shape and diameter of the exit nozzle (equivalent to the back edge of the jet tube), the rate of speed at which the gas leaves the exit orifice, the length of time of the "burn," and so on. With a little imagination, you can incorporate some of these factors into your match rocket experiments.

NEWTON'S LAWS, REVISITED

This experiment illustrates Sir Isaac Newton's third law of motion. When the match head is heated to ignition temperature, it forms a hot, rapidly expanding gas within the aluminum-wrapped match head (the rocket's combustion chamber). The rocket has two ports, or openings, that vent the gas in a controlled direction. The exiting gas propels the rocket forward. Newton's third law explains why: Any action has an equal and opposite reaction. In this case, if the gas is exerting a force in the backward direction (because that's the way our jet exhaust ports point), the equal and opposite reaction will cause the rocket to move forward.

ROBERT H. GODDARD

Professor Goddard with his 'chair' at Clark University…
does not know the relation of action and reaction;
[rockets] need to have something better than a vacuum against which to react. He seems to lack the
knowledge ladled out daily in high schools.

—*New York Times* editorial, 1921

Robert H. Goddard dreamed of launching a rocket through
the vacuum of space. The *New York Times* ridiculed him in
1921 for this "impossible" vision, but eventually Goddard
would be recognized as the father of the Space Age. In 1926,
while teaching classes at Clark University in Worcester, Mass-
achusetts, Goddard designed and built an oddly shaped contraption of steel tubes and nozzles. From a frozen farm field,
he launched the first rocket propelled by liquid fuels.

Most scientists thought that a rocket could not travel in
the vacuum of space, but Goddard's experiments proved otherwise. Even after he proved it possible, his ideas were
regarded with skepticism in the United States. Not so in Germany and Russia—"rocket clubs" and "experimental soci-
eties" popped up all over Europe, and the art of rocketry
advanced there. Ultimately, Wernher von Braun and other
Nazi scientists exploited Goddard's ideas and research to
produce V-2 rockets used during World War II.

The V-2 rocket was Germany's secret weapon and the first
ballistic missile ever developed and used in military applications. In 1944, from bases in occupied Europe, the V-2 rocket

flew higher than Allied fighter planes, faster than the speed of sound, and packed a huge, explosive wallop. These rockets wreaked havoc and destruction on England in the closing days of the war. Had the D-Day Allied invasion of Europe not taken place and the launch sites captured early on, the V-2 rocket could have changed the outcome of the war.

In 1969, the day after *Apollo* astronauts left Cape Kennedy for the first lunar landing, the editors of The *New York Times* published a public apology to Goddard, saying, "Further investigation and experimentation have confirmed the findings of Isaac Newton in the seventeenth century and it is now definitely established that a rocket can function in a vacuum as well as in an atmosphere." About the scathing criticism leveled at Goddard, the editorial added, "The *Times* regrets the error."

Unfortunately, Goddard never read the apology, and he never saw men walk on the moon. He didn't even get to see Russia launch the first satellite, *Sputnik*, which ushered in the Space Age in 1957. He had died of cancer 12 years earlier.

"Every vision is a joke," Goddard said, "until the first man accomplishes it." In honor of the memory of this rocketry pioneer, NASA established the Goddard Space Flight Center (GSFC) in 1959. GSFC takes a leading role in expanding our knowledge of space science. Scientists at GSFC design spacecrafts, build payloads for the space shuttle, and interpret much of the information collected through the U.S. space program.

The Hydro Pump Rocket: The Match Rocket's Big Brother

The principles at work in the match rocket can be applied on a larger scale to build the hydro pump rocket.

The hydro pump rocket is slightly more complicated than its smaller cousin and can be built in about two hours. It's a two-step process. First, you pressurize water inside a plastic soda bottle with a bicycle pump. Then, the pressurized water shoots out through a nozzle, propelling the rocket up into the sky. The hydro pump rocket will easily reach altitudes of 60 or 70 feet.

KEEPING SAFETY IN MIND

1. The rocket travels quite rapidly and would, as you can imagine, cause injury if it were to hit someone. So, it is very important to aim the rocket away from people.
2. Pressurizing the rocket strains the plastic, so inspect the plastic soda bottle after every launch and throw it away when signs of wear appear.
3. If the rocket lands on a hard surface and gets creased, throw it away and build a new water rocket.

BUILDING THE HYDRO PUMP ROCKET

The hydro pump rocket uses a stream of high-pressure water to provide forward thrust. The release of pressure shoots a stream of water through a nozzle and then the force propels the rocket into the sky.

It's impossible to predict the exact moment that the pressure inside the bottle overcomes the friction holding the rocket to the launch pad. This unpredictability adds to the fun, but be aware that this is an experiment where everybody gets wet.

This project modifies a two-liter soda bottle with fins and a rubber stopper. The stopper serves as a mechanism for pressurizing the rocket. We'll attach the whole contraption to a mounting platform so we can aim the rocket skyward.

Materials

- Electric drill with $\frac{1}{16}$-inch and $\frac{5}{32}$-inch drill bits

- (1) #4-size black rubber stopper (The stopper should be 1 inch in diameter at the top and 1 inch long. It should taper along its length as shown in the photograph on page 36.)

- (1) 8-inch length of $\frac{3}{16}$-inch outside diameter, soft copper tubing

- Standard inflation needle for inflating basketballs

- Foam rubber, or an old sponge 4 inches in diameter, shaped in a hemisphere, or a circle cut in half

- White glue

- (1) 2-liter plastic soda bottle (If the soda bottle has black plastic reinforcing on the bottom, remove it carefully without cutting into the bottle itself.)

- (4) 3-inch x 2-inch pieces of balsa wood

- Duct tape

- (1) plastic soda straw

- Wooden block approximately 6 inches x 6 inches x $\frac{1}{2}$ inch

- 2 wire clothes hangers

- Pencil

- Foot-stabilized air pump

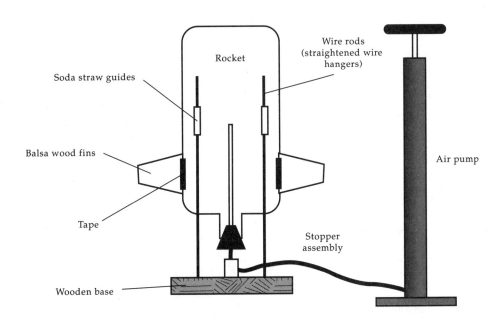

Making the Stopper Assembly

1. Carefully drill a $\frac{1}{16}$-inch hole through the middle of the stopper.
2. Next, widen the hole you just made by drilling a $\frac{5}{32}$-inch hole about $\frac{1}{2}$ inch down from the top (1-inch diameter) surface of the stopper.
3. Carefully straighten the 8-inch piece of copper tubing and insert it into the $\frac{5}{32}$-inch hole in the stopper as shown in diagram 3.9.
4. Insert the inflation needle into the bottom surface of the stopper so that it projects into the copper tube.

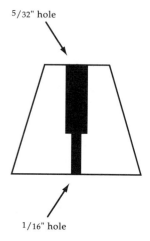

3.8 Water rocket stopper hole detail

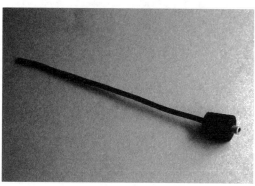
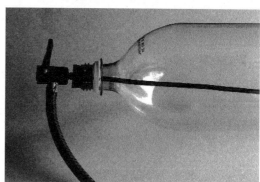

3.9, 3.10, 3.11, 3.12 Water rocket components

Making the Rocket

1. Cut out a half sphere of foam rubber (or an old sponge) and glue it to the bottom of the soda bottle (which is really the top of the rocket).

2. Attach the four 3-inch x 2-inch balsa wood fins to the rocket using duct tape or glue.

3. Attach three 1-inch lengths of plastic soda straw parallel to the fins, as shown in diagram 3.7.

Making the Launch Platform

1. Place the rocket on the wooden block and insert the straightened wire hangers into the attached soda straw guides so that they touch the wooden base. Mark the contact point with a pencil.

2. Drill a hole at each pencil mark on the wooden base and insert and glue the wire launching rods in place. The hole should be large enough so that the wire hanger fits snugly when inserted. Let the glue dry.

Launching the Rocket

1. Attach the bicycle pump to the inflation needle. (**Note:** See diagram 3.7.)

2. Fill the 2-liter bottle ⅓ full of water.

3. Push the rubber stopper assembly firmly onto the 2-liter bottle.

4. Invert the rocket and place the whole assembly on the wooden block launch platform by threading the soda straw guides onto the launching rods.

5. Firmly and slowly pump air into the rocket. The amount of pressure inside the rocket will vary depending on the cleanness of the seal between the rubber stopper and the rocket and how firmly the stopper was placed.

6. After a few pumps, the pressure inside the rocket will be

great enough to overcome the friction holding the stopper in place. Now for the fun part: The pressure will release the stopper from the rocket and the rocket will launch high into the air, shooting a thin trail of water behind it.

Remember, you might get wet!

The Pneumatic Missile

Thousands of years ago, primitive people in Asia and Europe used blowguns for hunting small animals. To use a blowgun, a hunter huffs into a long bamboo tube to shoot out a dart. Blowguns are very powerful, silent, and accurate. In fact,

3.13 Pneumatic missile

they are still used today by Amazonian tribes in South America and by Pygmies in Africa.

One of my friends recently showed me a toy he bought for his daughter. Like a blowgun, it uses a big push of air to launch a missile. It consists of a hollow plastic rocket fitted snugly over a tube attached to a large rubber bladder. When you jump on the bladder, the air pressure shoots the rocket high into the air. In this section, we expand on that idea and build a high-performance, homemade, pneumatic-powered missile.

BUILDING THE PNEUMATIC MISSILE

Pneumatics is the branch of mechanics that deals with the mechanical properties of gases, primarily air. So, a pneumatic missile is one that is propelled not by a chemical reaction, as in the two previous experiments, but instead by a rapidly expanding jet of air to push a simple rocket high into the air.

When you try this experiment, you'll be using a jet of compressed air to perform (sort of) useful work. Compressed air is commonly used in a wide variety of applications; pneumatic equipment powers items as diverse as construction tools, dental drills, and conveyor belts.

Materials

- (1) ½-inch diameter wooden dowel, 1½ inches long (This connects the base to the launch tube.)

- (1) 10-inch x 10-inch x ½-inch block of pine board or scrap wood

- Nails, screws, or glue

- (1) 12-inch length of ½-inch diameter schedule-40 PVC pipe

- (1) ½-inch x ½-inch x ½-inch PVC tee joint—two ends smooth, middle end threaded

- (1) ½-inch pipe to ½-inch tubing adapter (This is available at most hardware stores, in the department that sells plastic tubing.)

- (1) 18-inch length of ½-inch outside diameter, flexible plastic tubing

- (1) 28-ounce or larger plastic dish detergent bottle made from high-density polyethylene (Often, the bottles are labeled with the letters HDPE at the base. Remove the plastic nozzle, but retain the threaded part of the end cap.)

- File, knife, or drill to whittle dowel

- (1) ½-inch diameter wooden dowel, 1 inch long (This is the nose cone.)

- Scissors

- (1) piece of typing paper

- (1) ½-inch diameter wooden dowel 8 inches long (This is the mandrel, or form, for making the paper missile tube.)

Making the Launcher

1. Securely attach the 1½-inch long wooden dowel to the 10-inch x 10-inch wooden block by using nails, screws, or glue. The dowel should be perpendicular to the wooden block.
2. Insert the 12-inch PVC pipe into one end of the PVC tee joint. In the middle opening of the tee, screw the ½-inch pipe to the ½-inch tubing adapter. (See diagram 3.14.)
3. Attach the plastic tubing to the end of the tube adapter. This completes the launch tube assembly.

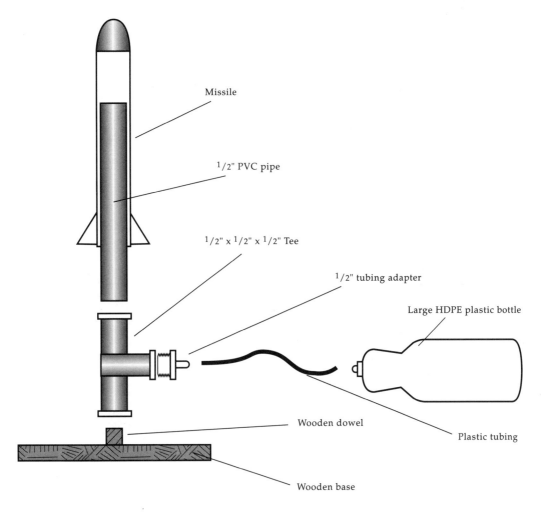

Missile

$1/2"$ PVC pipe

$1/2" \times 1/2" \times 1/2"$ Tee

$1/2"$ tubing adapter

Large HDPE plastic bottle

Wooden dowel

Plastic tubing

Wooden base

3.14 Pneumatic missile assembly

4. Push the bottom of this assembly onto the $1/2$-inch dowel attached to the base plate that you made in step 1 until the whole device is firmly attached in the upright position.

5. Stretch out the flexible plastic tubing and attach one end of it to the open end of the empty detergent bottle cap. It's a very tight fit; persevere, and it will eventually fit over the plastic tip. When the tube is attached, screw the cap back onto the plastic bottle.

Making the Missile

1. File or whittle the 1-inch wooden dowel into a rounded nose cone. Weight is very important in terms of the performance of your rocket, so make the dowel as light as possible by hollowing out the interior of the nose cone with a drill or knife. Drill a hole from the bottom of the cone, up toward the nose, but do not drill all the way through the nose.

2. Cut a sheet of paper approximately 3 inches wide by 6 inches long. Use the 8-inch long dowel as a mandrel to form a fairly loose paper tube 6 inches long. Use tape to hold the paper tube together.

3. Tape the top edge of the paper tube to the bottom of the nose cone. This forms the missile.

Launching the Rocket

1. Place the rocket over the launch tube.

2. With your foot, stamp quickly and firmly on the plastic bottle. The rush of air will send the missile flying skyward to a surprising altitude.

TIPS AND TROUBLESHOOTING

1. First, make sure all connections are tight and there are no places for air to escape except to push the rocket up.

2. Weight is of critical importance. Make the nose cone as light as possible for best results.

3. The bigger the air reservoir, the better the performance will be. Look for the biggest HDPE (high-density polyethylene) bottle you can find that will attach to the flexible plastic tubing.

ALTITUDE WITH ATTITUDE

Just how high will these rockets and missiles fly? You can't hold a tape measure up in the air and measure height directly, so is there another way to find out?

While it may be possible to compare the rocket's apex with some nearby tree or building of known height, that's inconvenient at best and impossible at worst. Luckily, there's a simple technique we can use to solve this problem—a technique known as "triangulation."

Triangulation is based on the branch of mathematics called trigonometry. Trigonometry allows you to determine the lengths of triangle sides when you know either the lengths of the other sides or the angles formed by the sides.

The first people to reason out the principles of trigonometry were the ancient Greeks, and the first known book on the subject was written by the Greek mathematician Hipparchus in about 140 B.C. Although his actual books haven't survived, it is believed that Hipparchus wrote twelve books on this and related mathematical subjects. Many writers and scholars claim that these books make Hipparchus the founder of trigonometry.

Let's determine how high our rocket flies using trigonometry. First, some background. Look at the accompanying diagram, which shows a right triangle. If you take the length of one side of the triangle and divide that length by the length of another side, you come up with a ratio. The ratios you get when you divide the sides of right triangles by one another are peculiar, and are the basis for trigonometry. Hipparchus and his fellow Greeks first figured out that the ratios of the sides of right triangles are related in very special ways, and that these special relationships could be used to determine a lot about any given triangle.

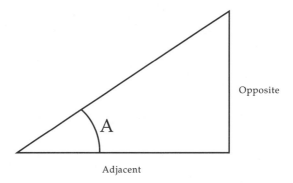

Opposite

A

Adjacent

3.15 Right triangle

Starting with the angle marked A, the ratio of the length of the side opposite angle A to the side next to (adjacent to) angle A is called the tangent. There are many other important ratios (sine, cosine, and so on), but we are only concerned here with the tangent.

The path that the pneumatic missile takes can be seen as one side of a triangle. (See diagram 3.15). What we're going to do is determine the tangent of this pneumatic missile triangle, and then work backward to determine the height that the missile shot up to when it was at its apex (its highest point in the sky). We need to know two things in order to calculate the height of our missile. First, we need to know the length of the side of this imaginary triangle that's adjacent to the angle we're going to measure. Then, when we measure the angle of the rocket at the apex, we can solve this equation:

$$\text{Tangent A} = \frac{\text{Length of opposite side}}{\text{Length of adjacent side}}$$

In other words,

$$\text{Tangent A} = \frac{\text{Height that missile reached at apex}}{\text{Length of ground from us to missile}}$$

We can rearrange this formula to read:

LENGTH OF OPPOSITE SIDE = (TANGENT A) X (LENGTH OF ADJACENT SIDE)

Because we can directly and accurately measure the length of the side of the triangle adjacent to the measured angle, we can easily calculate the height by looking up the tangent of the angle in a reference book and multiplying it by the known length.

To measure the angle the rocket makes with the ground, we need to make a protractor we can aim.

MAKING A PROTRACTOR

A protractor is a device that measures angles. In order to determine how high the rocket flies, we need to know what angle the sight line of the rocket at apex makes with the ground. You can make a protractor out of sturdy cardboard, a string, and the paper tube from a roll of paper towels.

Materials

- Cellophane tape
- (1) piece of cardboard, at least 8 inches x 10 inches
- (1) 12-inch piece of string
- Paper tube (Empty paper tube from a roll of paper towels works well.)
- Split-shot fishing weight

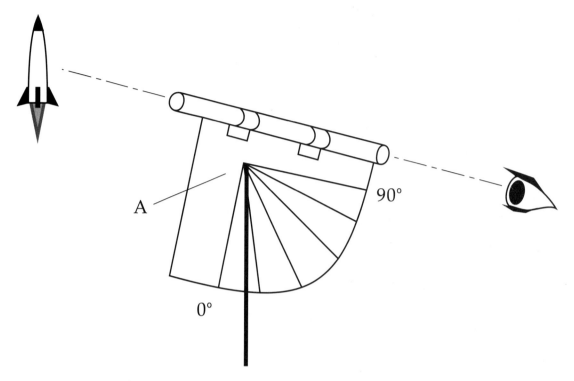

3.16 Making and using the protractor

Making the Protractor

1. Mark off angle lines on the cardboard from point A in 15-degree increments, starting from 0 degrees (level) and going to 90 degrees (pointing straight up).

2. Attach a string with a split-shot fishing weight at the bottom to point A. Then tape the tube to the cardboard as shown and you're ready to go.

3. When you launch the rocket, visually track the rocket through the paper tube and record the angle shown at apex.

To summarize, in order to figure out how high our missile goes, we can follow three easy steps.

1. Measure 100 feet in any direction from the launcher.
2. When your friend stomps on the plastic bottle and sends the rocket airborne, carefully measure the angle between the ground and the rocket's apex with the protractor.
3. Look up the angle in the attached chart. By using the tangent of the angle measured, and multiplying by 100 feet, you can calculate approximately how high the rocket went.

If the angle you measure is:	Then the rocket's top altitude is:
10 degrees	18 feet
20 degrees	36 feet
30 degrees	58 feet
40 degrees	84 feet
45 degrees	100 feet
50 degrees	119 feet
60 degrees	173 feet
70 degrees	274 feet

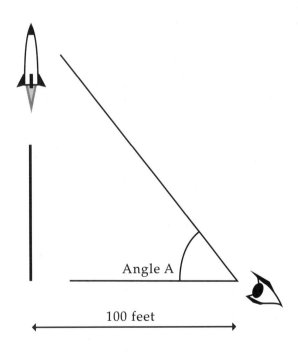

Angle A

100 feet

▷▷ 4 ◁◁

The Cincinnati Fire Kite

1946. A humid night on the Ohio–Kentucky border. A group of Fourth of July picnickers enjoys the warmth of the long summer evening.

Suddenly, someone in the group points to the south. There, behind a grove of small pine trees, a floating circle of fire rises like a phosphorescent Portuguese man-of-war in a dark lagoon. "What is that?" they wonder. "A UFO? A shooting star?" The Cincinnati fire kite makes its first recorded appearance.

Floating gently on the dense night air, the Cincinnati fire kite demonstrates the principles of buoyancy and lighter-than-air aeronautics. The fire kite is a simplified version of the original hot air balloon, which was first built by the French Montgolfier brothers around the time of the American Revolution. The difference is that the hot air balloon is heated by a separate burner, whereas the fire kite itself is the heat source, fuel, and containment device all in one simple package.

When you try this experiment, you'll be playing with fire. Read, study, and follow all the directions and notes regarding this project. Also, be aware there is a certain Zen quality to making a good fire kite. Sometimes, a kite will fly perfectly the first time. Other times, it will take many attempts before the kite flies the way you want it to.

Fortunately, newspaper is cheap, so keep at it. Once you get the hang of shaping a working fire kite, you'll find this to be a worthwhile and impressive experiment.

Building the Cincinnati Fire Kite

Pick the right time and place to launch your Cincinnati fire kite. If you don't, the neighbors may think UFOs are invading the city. (Honestly! I have seen the reactions of people when the fire kite floats above. They usually can't figure out what it is, leading to wild speculation.) When you perform this experiment successfully, you will send an ignited newspaper kite aloft for a short time. Carefully choose your spot for trying out this experiment.

As explained later in this chapter, the kite floats because the hot air inside it is less dense and more buoyant than the colder air outside it. Therefore, this project works much better on cool evenings than warm ones.

Mateials

- (1) full sheet (2 pages) of newspaper

- Stapler or transparent tape

- (4) books of matches or 4 long-handled fireplace matches

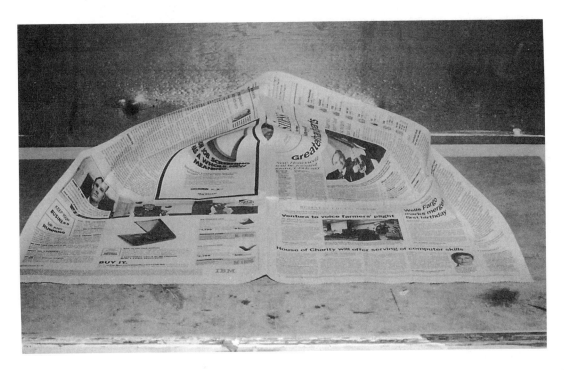

4.1 Folding the fire kite

1. Fold the newspaper as shown in diagram 4.1.
2. At the point where the four corners meet (point A, diagram 4.2), staple or tape the paper together to form the pillow-shaped kite. Adjust the kite so that all edges align as closely as possible. (It probably will not be possible to make all edges meet perfectly.) If you must, use bits of tape to fasten the edges.
3. Take the kite outside. It is now ready to launch.

4.2 Fire kite ready to fly   Point A

Launching the Fire Kite

1. Turn the kite over so the paper seams are on the bottom.
   With one, two, or three assistants, strike multiple
   matches and ignite the kite at each of the four corners.
   Try to ignite all four corners as close to the same instant
   as possible.

2. If all goes well, the kite will rise as it burns. It should
   ascend slowly. The charred kite will maintain its original
   shape and continue to rise and glow for several seconds
   longer. The effect is spectacular.

THE MONTGOLFIERS

In the late summer of 1783, the first hot air balloon arose from its launching site in the French countryside. Joseph-Michel and Jacques-Étienne Montgolfier had constructed a large fabric and paper bag and then placed a platform piled with burning straw and sticks beneath it. The bag inflated, expanded with hot air, and then rose into the sky to a height estimated by observers to be nearly 3,000 feet.

The Montgolfiers knew they were on to something big, so they redoubled their efforts and engineered larger balloons with greater lifting capacity. A few months later, the brothers sent up a sheep, a rooster, and a duck in a basket suspended from a balloon. While the bird's-eye view may have been old hat for the duck, it is likely the sheep was mighty impressed.

Since the farm animals—well, at least the sheep—seemed to enjoy the voyage, the brothers looked around for people adventurous (or crazy) enough to be human guinea pigs. Predictably, volunteers were hard to come by, so, according to one old story, the King of France stepped in. Louis XVI decreed that two condemned criminals must make the first flight. If they survived, a pardon would be their reward.

But others wanted the honor. In Paris, on November 21, 1783, two macho French noblemen, Pilatre de Rozier and the Marquis d'Arlandes, became the first humans to fly in a hot air balloon.

The Montgolfier brothers built their flying contraption out of paper and silk. The two noblemen, willing to risk life and limb for undying fame and glory, piloted the craft on a 22-minute flight from downtown Paris to a suburban vineyard several miles away.

Imagine what the local people must have thought when they looked up and saw a huge globe, shooting flames and hovering overhead. The farmers were nothing less than terrified of this fiery dragon descending from the sky. "Mon Dieu!" they cried, and charged it with shovels and pitchforks. Quickly, the balloonists explained their balloon to the farmers. The story verified, the noblemen passed a bottle of champagne to all present in celebration of a safe landing. Today, this custom is still observed by balloonists upon a successful landing.

TIPS AND TROUBLESHOOTING

1. Although extremely simple in design, the fire kite experiment can be difficult to perform successfully. It is necessary to practice and be persistent. It usually takes several tries before a good launch is achieved.

2. Cincinnati fire kites can reach heights of 30 feet or more before disintegrating at treetop level. Atmospheric conditions (temperature, humidity) play a big role in determining how fast and how high the kite rises. Remember, air is denser at cool temperatures, so launching is easier on a cool day.

3. It can be difficult to fold the kite so that the newspaper's edges align perfectly, but it is very important to get them as close together as you can.

4. Do everything you can to minimize the weight of the kite. Try to use a single staple or very small piece of tape at point A. Take time to readjust the paper or get a helper to hold the corners in place while the staple is placed precisely at the correct spot.

5. The hot air must be kept inside the kite in order to achieve the desired buoyancy. If needed, use a bit of tape to keep the kite as airtight as possible.

6. There is a tendency for the newspaper to burn unevenly. If this happens, the uneven burn opens holes in the kite and the hot air escapes too quickly. Try to synchronize the lighting of the ends so all flames reach the top at about the same time. Otherwise, the kite will flip over as it starts to rise, and the hot air will leak out of the open seams. If this happens, the kite doesn't rise.

KEEPING SAFETY IN MIND

1. The newspaper kite is set afire using matches. Be careful to avoid burns.

2. The kite will flame up just before it rises. Make sure that there is nothing flammable nearby in the air (trees, utility poles) or on the ground (dry grass, paper, matches, or other flammables).

CINCINNATI FIRE KITE PHYSICS

As the kite burns, it heats the air trapped in the interior of the kite. Hot air weighs less than cold air. The principle of buoyancy states that an object (the kite) completely immersed in a fluid (the cold air) will be acted on by an upward force equal to the weight of the fluid displaced minus the weight of the container (the newspaper and staple/tape).

In this case, the upward force is equal to the difference in weight between the hot air trapped in the kite and the cold air surrounding it. It's not a very large force, but it's greater than the force of gravity acting downward. The force is powerful enough to make the kite buoyant and rise up into the sky, glowing and floating like a small moon.

▶ IN THE SPOTLIGHT ◀

ARCHIMEDES OF SYRACUSE, HIERO'S HERO

In the third century B.C., King Hiero I of Syracuse challenged the great geometer Archimedes to solve a perplexing problem. The king had commissioned his royal goldsmith to create a new and irregularly shaped crown from pure gold. For reasons now unknown, the king was suspicious of the goldsmith. Did the goldsmith alloy a base metal, like silver or copper, to the gold to save money, and so defraud the king?

"Archimedes," said King Hiero, "I want you to determine the quality of this crown. I want to know if it is pure gold, or if it has been mixed with a lesser metal. However, you must not cut, nick, or deface my crown in any way. You must find out what the crown is made of without taking any samples or removing any material."

Archimedes loved a good scientific challenge. He figured all he needed to do was to determine the volume of the crown and then compare its weight with a similar volume of pure gold. But he couldn't directly measure the volume of the crown with a ruler because it was so irregularly shaped.

According to the often-repeated legend, the solution struck Archimedes one day as he sat down in his bath. The water in the tub spilled over the sides, and Archimedes recognized that for objects that did not float, the volume of water displaced is equal to the volume of the immersed object. Archimedes was so excited by this epiphany that he jumped out of the bathtub and ran naked through the streets of Syracuse shouting, "Eureka! Eureka!" ("I have found it!")

To determine the authenticity of the gold, Archimedes needed to determine the density of the crown. Density is a constant, nonchangeable property of an object determined by dividing the object's mass by its volume. Until the bathtub revelation, Archimedes had no way to measure the volume of the King's odd-shaped crown. However, by measuring the volume of the water displaced by the crown, he could learn its exact volume. He could then take a quantity of gold that was the same volume as that of the crown and compare its weight to the crown. If the weight was different, then the crown could not be pure gold and, hence, the goldsmith would be caught cheating.

This idea, called volumetric displacement, is the first step Archimedes made toward his initial understanding of the concept of buoyancy, which is often referred to as the Archimedes Principle.

In some ways, Archimedes was troubled by his fame. He considered working in the pure sciences and mathematics to be the highest level of scholarship, and his contributions there the crowning achievements of his life. Yet his contemporaries knew and loved him best for his engineering accomplishments—applications of science and math to real problems.

The people of Syracuse most admired Archimedes for designing the war machines that defended their walled city from the Roman Navy in 212 B.C. The Roman emperor, once an ally of King Hiero, had grown angry over Hiero's support of the rival city-state of Carthage. The Roman emperor dispatched his best admiral, Marcellus, with a large fleet to conquer the island fortress of Syracuse. Syracuse's army was puny compared to Rome's mighty legions, and they knew they wouldn't win by force alone. So they turned to Archimedes, now a frail and elderly mathematician, for inspiration and help.

When the war galleys of the Roman invaders drew near the walled fortress city, a fusillade of rocks and missiles rained down upon them, sinking their ships and panicking the legionaries. The Romans knew that no humans could be responsible for such an amazing attack—only the gods could have thrown boulders like that.

But in reality, the old geometer had designed and supervised construction of large catapults that had slung rocks from behind the fortress walls and down upon the enemy troop ships. The high trajectory of the rocks made it seem to the Romans as if the rocks were raining down from the heavens. They took this as a sign from the gods to disengage and flee.

Eventually, Marcellus and his navy returned and were able to breach the walls of Syracuse—by treachery and subterfuge, not by overcoming Archimedes' engines. According to the Roman historian Plutarch, after the Romans conquered Syracuse, Marcellus wanted to meet the famous mathematician of Syracuse and ordered his men to bring Archimedes before him. A Roman legionary found Archimedes working in his library. "Come now," said the Roman. "I must take you to Marcellus."

"Wait a while," replied Archimedes. "I am amidst work on an important problem, and the solution is near!" He refused to follow the Roman until he was finished. With a thrust of his sword, the impatient Roman made Archimedes pay for the delay—with his life.

After the Romans conquered Syracuse, the city, once the zenith of learning and scholarship, stagnated intellectually. The Romans concentrated wholly on the pragmatic and the practical. Whereas in Greece, the best and the brightest minds struggled with philosophy, theoretical mathematics, and pure science, in Rome, huge civil engineering projects—aqueducts, bridges, roads, and the like—dominated the best scientific minds.

Archimedes and the other pillars of Greek science and learning were sadly underappreciated. Cicero, one of Rome's best and most famous historians, wrote about the intellectual worldview of his contemporaries:

> Among them [the Greeks] geometry was held in highest honor; nothing was more glorious than mathematics. But we [the Romans] have limited the usefulness of this art to measuring and calculating.

Cicero understood the importance of the Greek contributions to pure science better than most. In 75 B.C., after studying the Greek influence on Roman civilization, he set out to find the grave of the long-dead mathematician. The people of Syracuse had no recollection of Archimedes or his contributions. In fact, most Syracusans, just 137 years after his death, had no concept of his contributions, or even that he existed! Through perseverance, research, and luck, Cicero came across the grave of Archimedes. He wrote:

I managed to track down his grave. The Syracusans knew nothing about it, and indeed denied that any such thing existed. But there it was, completely surrounded and hidden by bushes of brambles and thorns. I remembered having heard of some simple lines of verse that had been inscribed on his tomb, referring to a sphere and cylinder modeled in stone on top of the grave. And so I took a good look round all the numerous tombs that stand beside the Agrigentine Gate.

Finally, I noted a little column just visible above the scrub: It was surmounted by a sphere and a cylinder. I immediately said to the Syracusans, some of whose leading citizens were with me at the time, that I believed this was the very object I had been looking for. Men were sent in with sickles to clear the site, and when a path to the monument had been opened, we walked right up to it. And the verses were still visible, though approximately the second half of each line had been worn away.

So one of the most famous cities in the Greek world, and in former days a great centre of learning as well, would have remained in total ignorance of the tomb of the most brilliant citizen it had ever produced, had a man from Arpinum not come and pointed it out!

▷ 5 ◁

Greek Fire and the Catapult

After the fall of Rome, the empire eventually split into two parts. The western part was centered in Rome and the eastern part, called Byzantium, was ruled by a succession of emperors from the new capital of Constantinople.

Among the many capable Roman and Byzantine emperors, the emperor Justinian seems to have possessed particular vision and foresight and was especially successful in the use of technology to take and hold a military advantage. Military strength and national security had to be the utmost priority for Justinian, considering Byzantium's location—with uncivilized, raiding Germanic tribes to the west, acquisitive Russians to the north, and expansion-minded Turkish sultans to the east.

Justinian's commander in chief, Belisarius, was quite a military strategist and was best known for two great military innovations. The first was the development of armored horse cavalry, called the Cataphracti. Consisting of Greeks and mercenary allies, the Cataphracti were the world's best horsemen. They trained extensively in the use of the sword and lance for close-quarters fighting, and were masters of a specialized bow and arrow designed for use on horseback.

Time and again, they beat back foes of much greater size due to their superior riding skills and tactics as well as their huge, specially bred horses.

The other great Belisarian war invention was "Greek fire." As far back as 600 B.C., incendiary mixtures were used in warfare. Various concoctions were mixed together and hurled over the walls and gates of enemy fortifications. The ingredients in these mixtures included sulfur, pitch, sawdust, and oils of various types and weights. The formulation of incendiary compounds was, to this point, somewhat akin to making homemade chili—everyone had their own secret ingredients for making it better. Some ingredients made it more sticky, some more intense, and others harder to extinguish.

Some of these additives may have helped. For instance, adding sulfur gave the mixture a horrid stench. Other ingredients, such as salt, did not add much to the weapon's effectiveness. Salt made the compound burn with a bright orange glow, but did nothing else. Medieval recipes for incendiary compositions included such arcane and unusual ingredients as oil of benedict (made by soaking bricks in olive oil), saracolle (a tree resin collected only in Ethiopia), verjuice (juice from crab apples), and hot horse manure. The knights and nobles had their troops mix big batches of the flammable, foul-smelling gumbo in large wooden tubs. Then, they would fill thin wooden barrels with the compound, set the barrels afire, and hurl them at enemy ships and forts.

Of all the incendiary compositions, Greek fire stands apart from the rest. A frighteningly effective incendiary mixture, its secret formula was known only to the Byzantine high command. Unquenchable and unbelievably hot, the fiery substance shot from catapults made the fierce Hun and Mongol armies turn and run. The Greeks took such careful security precautions that the formula for the "wet, dark fire" never became known outside of Byzantium. When asked about it, Belisarius always replied that an angel had given the formula to Emperor Justinian as a gift.

With the help of this technology, the Byzantines fought off Russian and Arab sieges throughout the first millennium. Their incendiary liquid weapon was very effective: Of eight hundred Arab ships that attacked Constantinople in A.D. 714, fewer than one hundred returned home.

5.1 Trebuchet hurling cask of Greek fire

5.2 Trebuchet

Historians continue to speculate about the true composition of Greek fire. According to one account, it was made from petroleum (which seeped up naturally from the ground in present-day Iraq), bleached and ground animal bones, and quicklime. Although the formulation of Greek fire is beyond the capabilities of most modern amateur experimenters, the fabrication of a tabletop scale model of the catapults that hurled jars of this fearsome weapon makes for a very interesting project.

Catapults were the artillery pieces in use before cannons were invented. There were two types of catapults: spring-powered and gravity-powered. A gravity-powered catapult is often called a trebuchet and consists of a big bucket of rocks attached to one end of a long, levered pole. When released, the weight of the rocks causes the pole to hurl a projectile placed in a sling at the other end in a long arc at a target.

5.3 Onager

Spring-powered catapults, on the other hand, use the energy twisted into a compressed spring to fling rocks, arrows, and jars of incendiaries at enemies. The Byzantines favored a catapult design powered by a tightly twisted coil of horsehair rope. This type of catapult is known as an "onager." It is the catapult most often depicted on television and in the movies. (The term "onager" refers to a wild donkey that lived on the plains of the Middle East and had the habit of kicking rocks at pursuing enemies with its hind legs.)

BUILDING THE CATAPULT

In this chapter, we will build a model onager catapult. This design uses the torsion energy stored in a tightly wrapped cord bundle to spin a single, vertically mounted throwing arm to hurl a projectile. This type of catapult was developed around the first century A.D. and used as the workhorse of medieval artillery units through the time of the Crusades.

Materials

- Saw

- (2) $\frac{3}{4}$-inch x $\frac{3}{4}$-inch x 4-inch pieces of pine (upright supports)

- (1) $\frac{1}{4}$-inch drill, (1) $\frac{1}{8}$-inch drill

- (2) 1-inch x 1-inch x 4-inch pieces of pine (uprights)

- (2) 1-inch x 1-inch x 10-inch long pieces of pine (frame pieces)

- (3) 1-inch x 1-inch x 4-inch pieces of pine (cross members)

- (24) $1\frac{1}{2}$-inch long "box" nails or glue

- (4) 1-inch x 1-inch x 2-inch pieces of pine (footings)

- Small hook and eye

- (1) $\frac{5}{8}$-inch diameter wooden dowel, $8\frac{1}{2}$ inches long

- Ball-peen hammer

- (1) 2-inch diameter fender washer with narrow inside diameter

- (1) #10 bolt, $\frac{3}{4}$ inch long with lock washer and nut

- (1) 18-inch length of $\frac{1}{8}$-inch nylon cord

- (2) $\frac{3}{4}$-inch diameter washers

- (4) $\frac{1}{8}$-inch diameter dowel, each $\frac{1}{2}$ inch long

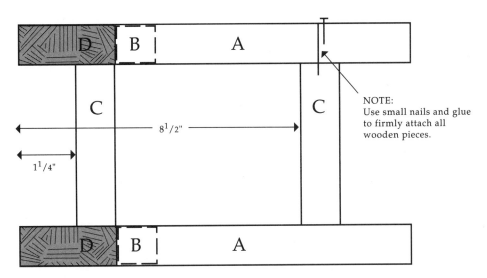

5.4 Top view of catapult assembly

NOTE:
Use small nails and glue
to firmly attach all
wooden pieces.

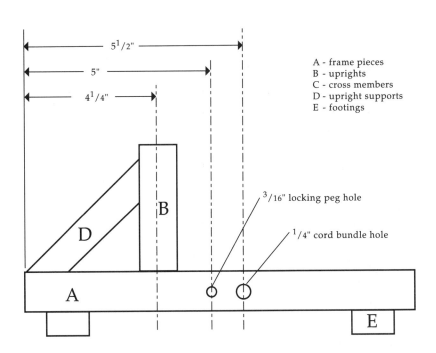

A - frame pieces
B - uprights
C - cross members
D - upright supports
E - footings

$^3/16$" locking peg hole

$^1/4$" cord bundle hole

5.5 Side view of catapult assembly

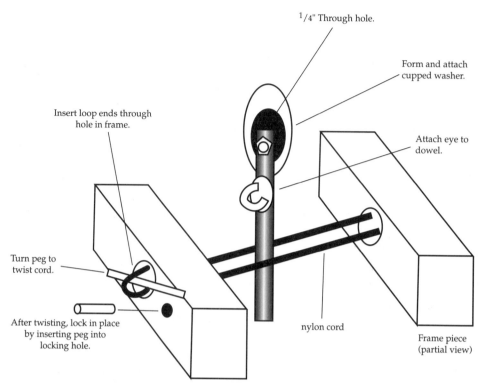

1/4" Through hole.

Form and attach
cupped washer.

Insert loop ends through
hole in frame.

Attach eye to
dowel.

Turn peg to
twist cord.

After twisting, lock in place
by inserting peg into
locking hole.

nylon cord

Frame piece
(partial view)

5.6 Catapult torsion spring detail

MAKING THE CATAPULT FRAME

1. Cut the ends of the pine pieces described as upright supports in the materials list at a 45-degree angle, as shown in the diagram.

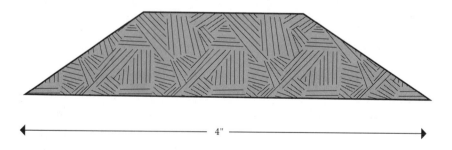

4"

5.7 Upright supports

2. Drill holes through the frame pieces in the locations shown in diagram.

3. Attach the uprights to the frame pieces as shown in diagram 5.5, using glue and/or nails.

4. Attach cross members to the framework as shown in diagram 5.5, using glue and/or nails.

5. Attach upright supports to the framework as shown in diagram 5.5, using glue and/or nails.

6. Attach all four footings to the framework as shown in diagram 5.5, using glue and/or nails.

7. Attach the hook and eye to the framework as shown in diagram 5.6.

Making the Throwing Arm

1. Screw the metal eye into the $\frac{5}{8}$-inch wooden dowel as shown in diagram 5.6.

2. Drill a $\frac{1}{8}$-inch hole in the throwing arm.

3. With the round end of the hammer, shape the fender washer into a cup.

4. Attach the washer, cupped side out, to the throwing arm with the bolt, the lock washer, and the nut.

Assembling the Torsion Spring

Refer to diagrams showing the torsion spring.

1. Tie the cord ends together securely with a square knot.

2. Hold the $\frac{3}{4}$-inch washers so that they rest on the outboard side of each $\frac{1}{4}$-inch diameter bundle hole. Insert the looped ends through each $\frac{1}{2}$-inch in the frame pieces and through the washers. Then, insert the $\frac{1}{8}$-inch dowel in each looped end of the cord that protrudes out of each washer-rimmed hole. Refer to the torsion spring diagram.

3. Insert the throwing arm through the cord and tighten the

cord by turning the throwing arm toward the front of the catapult over and over. Initially, you can slide the throwing arm up and down as you twist to avoid hitting the cross members. Once the cord is fairly tight, adjust the throwing arm so it is positioned as shown in the assembly drawing.

4. Complete the final tightening of the cord by twisting ⅛-inch dowels in each end of the cord bundle. Twist each tightening peg a few turns at a time, alternating sides. Be sure to twist the pegs toward the front of the catapult.

5. When tight, anchor the cord by inserting the stop pegs in the locking holes.

FIRING THE CATAPULT

1. Carefully pull the throwing arm back. Latch it with the hook and eye.

2. Place a projectile (such as a walnut) in the cup-shaped washer.

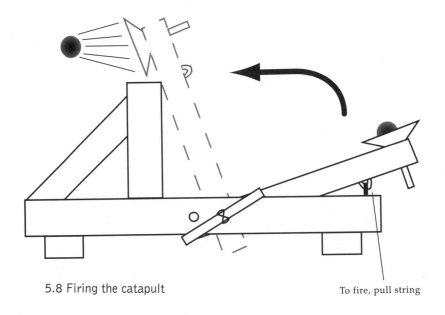

5.8 Firing the catapult

To fire, pull string

3. Attach a thin string to the hook. Grab the string and jerk the hook from the eye to fire the catapult. The more tension you apply to the torsion bundle (the twisted cord), the farther the catapult will shoot.

KEEPING SAFETY IN MIND

The throwing arm will strike the crossbar with sufficient force to squash objects trapped between it and the crosspiece when sprung. Watch your fingers!

▶ IN THE SPOTLIGHT ◀

NEWTON CALLS AGAIN

In 1687, Sir Isaac Newton published what may be the most influential book in the history of science, *Philosophiae Naturalis Principia Mathematica*. His book provided a framework for understanding how gravity affects the motion of all things—how fast things go, how far they go, how long they go, and so forth. In precise, highly technical Latin, Newton spelled out the rules for determining the motion of everything, from the apocryphal apple falling from a tree to the elliptical patterns the planets gracefully trace as they circle the sun. Newton's genius showed the universal application of gravity to all types of motion. It was Newton who first provided the knowledge that allows engineers and scientists to answer questions like "how high?", "how fast?", and "how long?" for almost any body in motion.

The catapult is a machine that imparts a force to an object in order to make the object fly until gravity pulls it down. As the cord (or as medieval catapultists would say, the "torsion bundle") is twisted, the energy imparted by twisting the turning pegs is transferred and stored in the fibers of the bundle. There, the energy sits until freed when the hook release mechanism is tripped.

After the catapult is fired, the potential energy stored in the torsion bundle is converted into kinetic energy, spinning the firing arm and hurling the projectile in a high parabolic arc toward its target.

Newton and his contemporaries—Robert Hooke, Jean Picard, and Gottfried Wilhelm Leibnitz—were the first scientists to develop the scientific theory and mathematics to allow modern engineers to completely describe the physics of a catapult. Many of the scientific terms we take for granted today—energy, force, power, acceleration, momentum, and so forth—were first developed by Newton and his fellow scientists.

The Catapults of History

DATE	PERSON	FACT
400 B.C.		Earliest record of a gravity-powered rock-throwing device (called a trebuchet) in China.
399 B.C.	Dionysius of Syracuse	Greeks build a giant bow called a gastraphetes. This was the earliest known nontorsion catapult. It was built to hurl darts.
397 B.C.	Dionysus of Syracuse	First decisive use of catapults in warfare. Greek forces use catapult-fired bolts to defeat the soldiers of the Cathaginian city of Motya.
345 B.C.	Phillip of Macedonia	First mention of torsion-powered arrow-shooting artillery.
332 B.C.	Alexander the Great	Catapult dart seriously injures Alexander.
330 B.C.	Alexander the Great	Engineers working for Alexander the Great, son of Philip II, build first torsion-powered stone throwing catapults.
305 B.C.	Demetrius	Demetrius of Greece invents huge movable towers on wheels. These towers carry scores of catapults and are used in sieges of walled cities.
212 B.C.	Archimedes	Archimedes of Syracuse builds powerful war engines to hold off Roman invaders under the command of General Marcellus in the siege of Syracuse.

DATE	PERSON	FACT
200 B.C.		Torsion-powered catapults supplant bow-powered devices.
A.D. 50	Julius Caesar	Torsion engines in widespread use in Roman army.
70	Titus	Romans under General Titus capture Jerusalem using hundreds of catapults against the city's inhabitants.
500		Earliest record of gravity-powered trebuchets in use in the Middle East.
600		Trebuchet artillery technology reaches Europe.
675	Justinian of Byzantium	Byzantines invent Greek fire, a powerful incendiary, for use in catapult warfare.
886	Paris	Nontorsion bow-powered catapults reintroduced and used in warfare with Gauls.
1191	Richard I	Franks and Turks batter each other with over 300 catapults in the Siege of Acre during the Third Crusade.

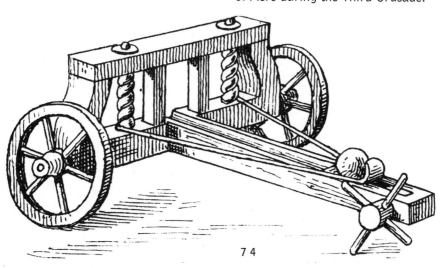

DATE	PERSON	FACT
1204	Philip Augustus of France	Cabulus, a huge trebuchet, is used by the French to hammer down walls of Chateau Galliard during a seige and chase the English King Henry III from Normandy.
1340	John, Duke of Normandy	The English and French of Normandy barrage each other with rocks and bolts as trebuchet warfare reaches its zenith in Europe.
1450		Cannon supplants catapult throughout Europe.
1480		Last successful use of catapults (trebuchet) in warfare during the battle of Rhodes.

▷ 6 ◁

The Tennis Ball Mortar

What happens when you duct-tape a couple of potato-chip tubes together, then add a source of energy, a tennis ball, and a match? Well, not much—unless you know the secret of building the fabled tennis ball mortar.

This chapter describes the design and construction of this nifty backyard boomer project. What's a mortar? How is it different from a cannon? Sometimes the words describing shooting devices are used interchangeably when they shouldn't be. There are differences among cannons, mortars, and howitzers.

> ▷ A mortar is loaded with ammunition through its muzzle— that is, the shell is dropped in the barrel of the device and allowed to fall to the breech, or back end. It generally shoots a shell at a low velocity and at a high angle of fire, or trajectory.

- ▷ A howitzer shoots at a higher velocity and at medium-to-low trajectories.
- ▷ A cannon or gun is a more generic term and refers to a device that loads through its breech and fires a projectile in a flat trajectory and at high velocity.

The device discussed in this chapter lobs a tennis ball a short distance, but in a very high, arcing trajectory—ergo, the tennis ball mortar.

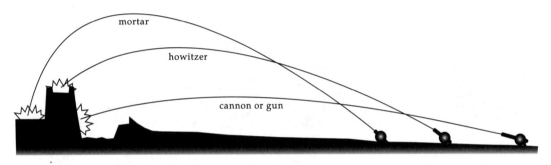

6.1 Artillery trajectories

Building the Tennis Ball Mortar

Lighter fluid supplies the energy for the tennis ball mortar. The real name for lighter fluid is naphtha, and it is flammable and therefore dangerous. Kerosene, naphtha, heating oil, and so forth are petroleum distillates produced by processing crude oil through a refining process. Each distillate has a substantially different level of energy and volatility (a substance's tendency to evaporate).

Follow the directions carefully, and don't use more lighter fluid than specified. Of course, don't use any flammable material other than lighter fluid. High volatility, high-energy liquids such as gasoline are very different in their burn rates and volatility and therefore will not work according to the steps outlined in this chapter. *Do not use any substances other than those described here.*

Materials

- (3) cardboard tubes or cans approximately 10 inches long by 2½ inches in diameter (**Note:** The cans in which some processed potato chips are packaged work well because they are the ideal diameter for launching a tennis ball. Further, they are coated with a grease-resistant polymer that won't readily soak up the propellant fluid.)

- ³⁄₁₆-inch diameter drill or punch

- Tin snips

- File or sandpaper

- Duct tape

- Thin-walled, 4-inch diameter PVC sewer pipe, length equal to the length of number of cardboard tubes, end to end (This forms the blast tube and provides protection.)

- Lighter fluid (Lighter fluid is available in containers with a plastic pouring tip. Butane lighter fuels are not suitable for this experiment.)

- Tennis ball

- Wooden stick, about 12 inches long and thin enough to fit into the mortar

- Protective gear including earplugs, gloves, and safety glasses (The mortar can be very loud. Everyone nearby must wear hearing protection.)

- Several bricks or large stones

- Long-handled lighter or fireplace match

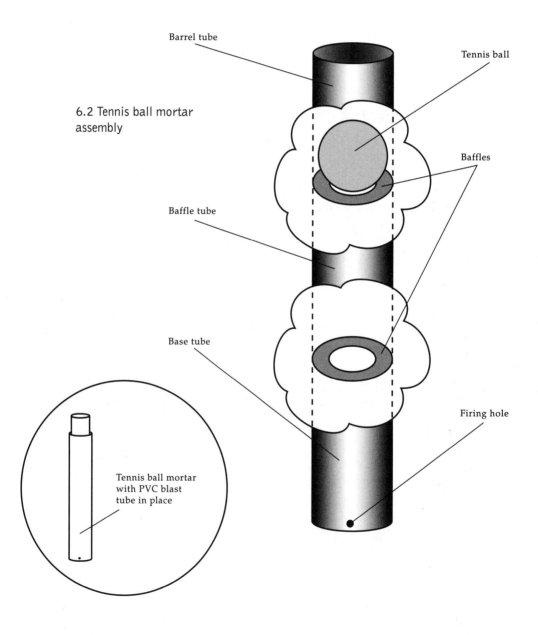

6.2 Tennis ball mortar assembly

Barrel tube

Tennis ball

Baffles

Baffle tube

Base tube

Firing hole

Tennis ball mortar with PVC blast tube in place

Making the Mortar

1. Choose one empty tube to be the "base tube." On this tube, poke or drill a clean $\frac{3}{16}$-inch hole approximately 1 inch up from the bottom.

2. On the two nonbase tubes, perforate the bottom end with a drill or tin snips so you have one round-shaped hole in the end. The amount of open space in the bottom of each tube should be about 50 percent of the total area. If you're using a $2\frac{1}{2}$-inch diameter tube, make the baffle hole about $1\frac{1}{2}$ inches in diameter. The $1\frac{1}{2}$-inch diameter cutout should be concentric (see diagram 6.2). Baffles are the key to good mortar performance. Make your baffles carefully.

3. File or sand the edge of the newly cut holes to remove burrs and dull them so they are safe to handle.

4. Assemble the mortar as shown in diagram 6.2.

5. Cover the entire assembly with no fewer than three heavy layers of overlapping duct tape. Make the seams between the cans completely airtight. If you have trouble with leaks, you can improve the seal between cans by applying caulk or glue to the can seams, letting it dry, and then taping it several times. Be sure to reinforce the back end cap joint (where the round metal end joins the cardboard tube) with tape. Be sure to leave the firing hole open.

6. Make a PVC blast tube out of the 4-inch PVC sewer pipe. The PVC tube should be made just long enough to contain the entire tennis ball mortar when it is slipped into it.

7. Insert the taped tennis ball mortar into the PVC blast tube. Drill a large hole in the blast tube that corresponds to the location of the firing hole.

Preparing the Tennis Ball Mortar for Operation

1. Use no more than $\frac{1}{2}$ teaspoon of lighter fluid. Drip the lighter fluid down the muzzle, making sure at least some

of it flows all the way down to the bottom can. **Note:** The baffles can make it difficult to get the fluid into the bottom firing tube.)

2. Place the tennis ball into the mortar. The ball will roll down to the first baffle and rest up against it. As it does, it keeps the fuel-air mixture inside the combustion area of the mortar. Take the wooden stick and press it against the ball so it holds it in place against the baffle.

3. Shake the mortar so the fluid is evenly vaporized throughout the inside of the device. Wait 20 seconds for the vapor to permeate the entire chamber. **Remember:** There is never a need to use more than ½ teaspoon of lighter fluid. Using more than this may be dangerous.

4. Move the mortar into a firing position with a high firing angle. Remember to aim the device carefully because the tennis ball will exit the mortar at high velocity and could injure someone.

FIRING THE MORTAR

1. The mortar makes a very loud noise. Everyone nearby must wear earplugs and safety glasses. The person firing the mortar must also wear gloves.

2. Prop the mortar into a high-firing angle using bricks or large stones. *Do not hold the mortar in your hands.* The mortar must be on the ground.

3. Align the firing hole in the PVC pipe with the firing hole on the mortar. Strike a very long match, such as a fireplace match, or use a long-handled piezoelectric lighter, and bring it to the firing hole.

4. The naphtha will ignite immediately and propel the tennis ball 20, 30, or 40 feet high or more. The loud report adds to the effect.

6.3 Mortar in firing position without the blast tube

TIPS AND TROUBLESHOOTING

1. The mortar can be two to three cans in length. (A two-can mortar is a good starting size.) Place a baffle between every can.
2. Make your baffles carefully—they are the key to good mortar design.

THE ROMAN CANDLE

In ancient times, people knew that if they added sulfur to charcoal, the resulting mixture would burn more rapidly and vigorously than just charcoal alone. As early as A.D. 1044, the military scientists of the Chinese Jin dynasty started to add a third compound—saltpeter—into the mixture, which made burning even faster.

What prompted the Chinese to add saltpeter to the mixture? Chinese engineers knew that adding table salt—sodium chloride—to the mixture of charcoal and sulfur made it burn with a bright orange flame. Engineers then, as today, were always seeking to improve performance, so they started to experiment with other types of salt. Eventually, they tried another common type of salt—potassium nitrate, or saltpeter.

Over time, the methods of combining the three ingredients became more refined. Through trial and error, the proportion of about one part sulfur to one part charcoal to four parts saltpeter was found to give the best results. Ground into fine powder and packed tightly into a closed container, these ingredients could explode with a blast of surprising power. This compound of sulfur, charcoal, and saltpeter became known simply as black powder.

The Chinese experimented with the military applications of explosive powders by packing the powder into bamboo tubes and layering it with rock: one layer of powder, followed by a layer of rock, another layer of powder, rock, and so forth. This mortar-like device was the first Roman candle.

As the burn proceeded down the muzzle, the layers of rock shot out of the bamboo tube one at time, timed by the distance between layers. This device propelled solid hunks of rock in rapid succession.

Soldiers found the Roman candle to be a useful part of their arsenal. Although the candles were not particularly dangerous to opposing troops (compared to arrows or slings), they were good at spooking the enemy's horses and terrifying superstitious enemies.

One question that stumps scientists and lexicographers alike is the name—Roman candle. If the Chinese invented it, why isn't it called the Chinese candle?

3. Reinforce all joints with duct tape.
4. Carefully ensure that the lighter fluid is fully and evenly dispersed throughout the device.
5. Use the blast tube.
6. It can be difficult to get the ignition flame properly positioned at the touch hole. Make sure the hole is free from tape and tape residue.
7. If the mortar fails to ignite, remove all ignition sources from the area. Carefully invert the mortar so the tennis ball falls out. Allow the mortar to air out completely and start again.
8. After you've had some practice, consider a night firing. Look for a large blue flame at the muzzle.
9. In scientific terms, firing the cannon results in initiating a vigorous exothermic (energy-releasing) reaction in the barrel of the mortar. The air/lighter fluid mixture is converted to a mixture of gases called "products of combustion" and energy. After firing, the products of combustion

will linger in the cannon and prohibit the cannon from firing again until they are cleared. This process takes a little while, and it is often necessary to let the device air out for a while between firings. Baffling makes the evacuation process even lengthier. You can use a hair dryer or shop vacuum cleaner to speed the evacuation process.

10. The life of a cannon is, at best, about three to five successful firings. Each tennis ball launch can put stress on the baffles, push out the end cap, separate the tube joints, and so on. Examine the tube after every firing and discard it if it starts to bulge or rip, or becomes noticeably worn out.

KEEPING SAFETY IN MIND

1. Firmly stabilize the mortar with a heavy base such as bricks or rocks.
2. Use only the recommended type and amount of fuel.
3. *Always* aim away from buildings and people.
4. *Always* use gloves and hearing and eye protection.
5. Examine the mortar after each shot. Discard the mortar when it is worn out.

INSIDE THE BARREL OF A TENNIS BALL MORTAR

Ballistics is the branch of physics that deals with motion of projectiles and the conditions that govern that motion. Sometimes ballistics is called the "science of shooting." The study of ballistics is divided into two areas: interior and exterior ballistics. Exterior ballistics deals with projectiles not under propulsive power; for instance, it describes what goes on after the tennis ball leaves the muzzle of the mortar. Interior ballistics is concerned with describing and understanding the explosive process that takes place within the barrel of a gun.

Inside the mortar, the fuel (i.e., lighter fluid) mixes with air in the barrel and forms a fuel-air mixture. This mixture will undergo a very rapid chemical reaction when put into contact with a flame. The chemical reaction, or burn, exerts a force on the entire interior of the tube assembly. However, the walls and bottom are rigid, so the force is channeled against the tennis ball and propels it out of the tube at high velocity.

As mentioned earlier, interior baffles are the key to good tennis ball mortar performance. Why would the presence of the baffles result in such a large increase in the distance the ball travels? Well, the baffles add a great deal of complexity to the analysis of the interior ballistics within the cannon. The computations, measurements, and analysis required to determine exactly what role the baffles play in this mortar have not been performed under laboratory conditions, but there are several possible explanations of their role. First, the baffles momentarily raise the pressure inside the cannon by restricting the flow of gas. When hot gas hits the baffle, it has to move across an area with a smaller cross section than the rest of the mortar. As the rush of gas passes the area of restriction, the pressure (and therefore the gas velocity) may go up. A higher gas velocity produces more force on the ball.

A second possibility is that the baffles amplify the force of the reaction by slightly slowing down the burn rate, allowing a more complete burn and obtaining more energy from the fuel-air mixture.

▷ 7 ◁

The Flinger

Sir Isaac Newton's stature in the world of science and engineering is second to none. He is universally recognized as one of history's most important physicists. But Newton was, at least in some ways, not a particularly pleasant person. His contemporaries found him reclusive, egotistical, and relentless in his endeavors. When you consider his intellect, you may forgive him for the pride that engendered many of his faults. And he wasn't completely unapproachable as a person or a scientist, for he did have close friends and shared credit for discoveries when shared credit was due.

Sometimes freely, and sometimes grudgingly, he acknowledged the help and contributions of other scientists. Newton once said, "If I have seen further, it is by standing on ye shoulders of giants." One of those giants was Robert Hooke, a man of towering intellect and diminutive physical stature who rivaled Renaissance man Galileo in the breadth of his scientific contributions. Hooke's work fostered important advances in the fields of physics, architecture, optics, engineering, and biology.

To say that Newton did not appreciate Hooke or his impressive body of scientific work would be quite an understatement. Newton recognized Hooke's genius, even though he and Hooke hated each other. Their feud started when Hooke published comments questioning one of Newton's treatises on optics. Newton's huge ego made him unable to accept the criticism. One bitter remark led to another, and soon the two scientists detested one another in a manner rarely seen before or since in the collegial halls of English academia. Given Newton's immense stature in London's scientific community, being on his bad side would have extinguished the career of a lesser light. But Robert Hooke's ability and discoveries in so many scientific fields permitted his genius to shine through Newton's cloaking of his professional contributions.

Hooke and Newton's arguments, shouting matches, and exchange of acrid letters took place in the halls of a prestigious scientific organization called the Royal Society of London. Here, the most highly esteemed seventeenth-century scientists published, discussed, and often argued their theories and declamations. During the years of their mutual Society membership, these men crossed swords on just about every topic—dynamics, optics, astronomy, celestial mechanics, and others.

Newton outlived Hooke, and he wasn't the type of fellow to let bygones be bygones. In a colossal fit of professional pique and petty jealousy, he did everything he could to expunge the late Hooke's name and image from the historical records of the Society. He was to some degree successful—all portraits of Hooke, his letters, and the apparatus with which he performed his great experiments were suspiciously "lost" by the Society, and for this many scholars blame Newton.

Despite this, Hooke is recognized today as an important contributor to many fields. He even has a basic law of physics named after him, Hooke's Law. It was derived from his theoretical work on potential and kinetic energy and summarizes how mechanical springs work. Hooke's Law forms the scientific basis for this chapter's project, the flinger.

Building the Flinger

The flinger is easy to make and fun to use. It is made of a loop of elastic rubber tubing stretched between two handles. To use it, a water balloon or similar object is placed in the pouch. The pouch is drawn back, released, and whoosh!—the balloon flies downfield. This is a relatively easy project, as the whole device is made up of only a few parts.

However, pay special attention to safety. Don't overextend the tube and don't aim at people. Hitting a person with a water balloon can be dangerous.

Materials

- (4) ¾-inch diameter schedule-40 PVC-pipe elbows
- (2) 3½-inch-long lengths of ¾-inch diameter PVC pipe
- PVC primer
- (1) can PVC cement

- (1) 1-foot length of strong 1-inch nylon webbing, sewn into a loop

- (1) 6-inch x 6-inch rectangle of nylon cloth or webbing for the pouch

- (1) 14-foot length of $^7/_{16}$-inch diameter, $^1/_4$-inch bore elastic rubber tubing (Tubing like this is available at hardware stores, large home stores, and medical supply companies. It is generally sold by the foot.)

- Electrical tape

- Water balloons

- Safety glasses and gloves

- Sewing machine

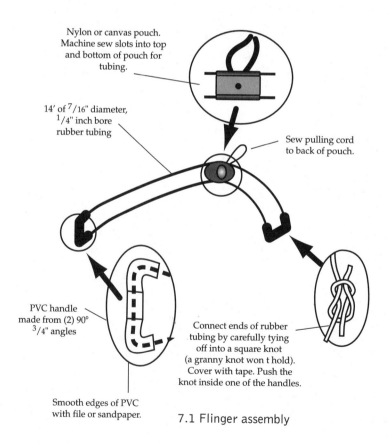

Nylon or canvas pouch. Machine sew slots into top and bottom of pouch for tubing.

14' of $^7/_{16}$" diameter, $^1/_4$" inch bore rubber tubing

Sew pulling cord to back of pouch.

PVC handle made from (2) 90° $^3/_4$" angles

Connect ends of rubber tubing by carefully tying off into a square knot (a granny knot won t hold). Cover with tape. Push the knot inside one of the handles.

Smooth edges of PVC with file or sandpaper.

7.1 Flinger assembly

Making the Handles

1. Assemble the PVC-pipe elbows and pipe lengths into two handles as shown in diagram 7.1. Solvent-weld all connections using PVC cement, following all directions on the can, including cure times.

Making the Balloon Pouch

1. Fold over the long edge on each side of the nylon and machine sew the fabric so that it forms a large enough slot to accommodate the rubber tubing. Use strong thread and plenty of stitches per inch. Put a slot on the top and on the bottom.
2. With a sewing machine, securely attach the nylon loop strap to the back of the pouch, exactly in the center. Again, use strong thread and several stitches per inch.
3. Insert the rubber tubing into the fabric slots.

Final Assembly

1. Tie the ends of the tubing together using a square knot, with at least 1 inch of tubing extending beyond the knot. It is very important that the knot you use be a secure one, like a square knot, and that you tie it so that it doesn't come loose. Once the knot is tied, tape it so that the ends hold securely.
2. Insert the rubber tubing into the other components (see diagram 7.1).

KEEPING SAFETY IN MIND

1. Check the rubber tubing for nicks and wear each time you shoot. Make sure the knot is tight and secure. If the flinger looks worn, replace or repair the worn parts before using it again.

2. Wear safety glasses and gloves for protection in the unlikely event the tubing breaks.

3. Don't aim at people. Make sure the area downrange is clear of people and other hazards. Keep nonparticipants out of the area in which you're working.

4. Don't overextend the rubber tubing. The people holding the handles should not be more than 6 feet apart. Pulling too hard on the pouch may rupture the tubing or hurt somebody. Limit the pull to about 30 to 35 pounds of force. **Remember:** Don't pull too hard or extend the rubber tubing.

5. Keep the pouch centered in the tubing, and don't launch hard, heavy, or dangerous items.

OPERATING THE FLINGER

1. It takes three people to operate the flinger. First, determine the direction in which you want to launch the water balloon. The flinger is not a particularly accurate device, so make sure the firing range is large, open, and clear of hazards.

2. One person grasps each of the PVC handles. These two people walk in opposite directions until they are about 6 feet apart. The tubing will droop initially, but that's OK. It is very important to hold the handles securely. Always wear gloves and eye protection.

3. The third person places an object such as a water balloon in the pouch.

4. Next, the third person pulls back on the rope. The tubing stretches, and as it does, it stores potential energy. The three people should form a *V* measuring less than 30 degrees.

5. The third person releases the rope and lets the tubing spring back. The balloon flies out of the pouch.

FLINGER PHYSICS

The flinger uses the energy stored by the elastic tubing to hurl the balloon. The tubing is not stretched in a line parallel to the direction of fire. Instead, the tubing forms a *V* when the pouch is pulled back, which makes figuring out the physics of the flinger a little complicated. When you pull back on the strap attached to the pouch of the elastic flinger, you are applying a force that works in a direction 180 degrees from the direction you want the balloon to go. Because the flinger, as a system, forms a triangle, there are forces acting in two directions.

First, you're pulling the elastic to your chest, straight back. Second, you're pulling on the sides of the elastic, making them longer, stretching them as you pull the middle of the band toward your chest. When you release the elastic tube, the forces move in two ways: (1) the tubing flies forward, straight away from you—the direction the balloon flies, and (2) the elastic snaps back to its original length (from being stretched all the way back). That force, the shortening of the tubing, is perpendicular to the flinger's line of fire.

"Vector" is a term used to describe physical qualities that have both an amount and a direction. Think of velocity as an example. To an engineer or physicist, a car doesn't just go 50 miles per hour. It goes 50 miles per hour east; velocity is speed and direction.

Like velocity, forces are vector quantities—they have an amount (an amount of energy such as 10 pounds-force, 25 newtons, 100 kilograms-force, etc.) and they have a direction. There are backward forces, sideways forces, upward forces, and downward forces. Therefore, engineers always think of forces as vectors.

As you pull back the flinger's cord, it is important to note that a wide, flat triangle (see diagram 7.2) puts most of the force on your tubing in the perpendicular direction. That's bad; you put a lot of strain on the rubber tubing and get little in return. A 45-degree angle puts half the force in the right direction and wastes the other half. For the best performance, make the angle nice and tight, and most of the force vector will act in the right direction—and the water balloon will go a long way.

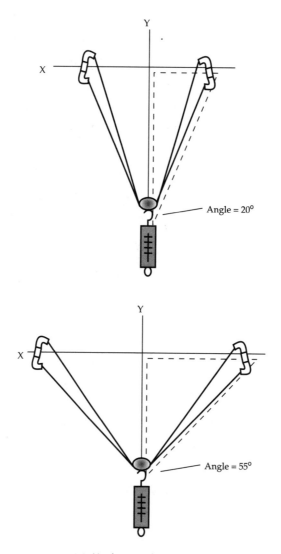

7.2 Understanding vectors

When you pull back on the rope, how much force is applied by the stretching of the tubing to the balloon? Robert Hooke was the first person to describe and predict what happens in precise technical terms. He determined that the amount of force it takes to stretch a spring is directly proportional to the distance it is displaced. This relationship is normally expressed as

$$\text{Force} = \text{Spring Constant} \times \text{Deflection},$$
$$\text{or } F = k * D$$

The flinger's rubber tubing is actually a type of spring. It stretches in direct proportion to the amount of force you place on it. The more force you apply, the more it stretches, and this amount is the displacement. When you let go of the pouch, the amount of force acting on the water balloon at that instant is equal to the amount of force it took for you to pull the pouch back. If you were to measure the force holding back the pouch when the tube was stretched one foot, then two feet, then four feet, you'd find the ratio of force to the amount of stretch (displacement) would be a constant number. This "linear relationship" is Hooke's Law, and it allows engineers and scientists to mathematically calculate how far the balloon will travel at various amounts of stretch.

THE MEDIEVAL CROSSBOW

The surgical rubber tubing in the flinger is a spring, although it doesn't look like the coiled, tempered wire that most people think of when they envision springs. Actually, springs come in many forms: coil springs, leaf springs, axial springs, torsion springs, and so on. The bows used in both the English longbow and the Genovese crossbow are actually springs, too.

Back in the fourteenth century, the state-of-the-art weapon for warring foot soldiers wasn't the sword, the pike, or the halberd. It was the bow and arrow. At the time, the crossbow was considered the most advanced weapon available. It was easy to aim, simple to use, and did terrible damage. It was made out of a composite spring (the bow) consisting of multiple layers of animal ligaments and wood fibers, all laminated together with glue made from a fish called the river sturgeon. The crossbowman would crank back the bowstring using a ratcheted wheel called a cranequin, lay in a short but heavy arrow (called a "bolt"), and fire it like a sharpshooter fires a rifle today.

Learning to use a crossbow was quick and easy—too easy in the minds of some government and religious officials. With a few days of practice, any soldier could aim and fire a crossbow. The longbow, by contrast, required that the soldier have years of practice and a very strong set of arm muscles to approach proficiency. The ease in which one could become proficient with a crossbow led to the first enactment of "gun control" laws. In 1139, Pope Innocent III issued a communication called a papal bull that outlawed the use of crossbows by all Christians. They were, he said, too easy to use, too powerful, and too deadly. He remarked that if weapons like this got into the hands of insurgents or heretics, well, that could shake the foundations of government and religion. Fearful of the destabilizing potential of the crossbow, the Christian countries of Europe obeyed Pope Innocent's proclamation and destroyed their crossbows.

This was the time of the Crusades; however, the English and Frankish armies soon found themselves in pitched battles against Turkish troops who had no such proscription against the use of crossbows. The fusillade of crossbow bolts shot by Turkish defenders made the Crusaders reconsider using the deadly and effective crossbow. Eventually, Richard I of England, better known as Richard the Lionhearted, reintroduced crossbow use among the rank and file of Crusader troops. Later, as it turned out, he would probably wish he hadn't.

As English kings go, Richard the Lionhearted, of Robin Hood fame, wasn't a very competent administrator or even a particularly smart man. He did only one thing very well—he was great at laying siege to castles. If there was a castle around, chances are Richard would find a reason to besiege it. From fighting the Turks at Acre to his conquests of Norman cities in France, Richard traveled throughout Europe and Asia in pursuit of a good siege.

He was brave, but in the end, foolishly so. During his 1199 siege of Chaluz Castle in Normandy, he deliberately exposed himself to enemy crossbow fire, presumably for no better reason than to show off his kingly courage. One of the castle defenders, Bertrand de Gourdon, saw Richard move within easy range of his crossbow and buried the bolt into

Richard's neck. Thus ended the reign of Richard I. Richard may have had the heart of a lion, but he seemed to have the brain of a bird.

About one hundred years later, the French and the English were still warring. In 1415, 6,000 longbow archers of Henry V of England met and fought a force of 25,000 French knights, pikemen, and mercenaries at the Battle of Agincourt, in France. The French-led mercenaries were Genovese soldiers of fortune, and they were equipped with state-of-the-art Italian crossbows. On the other hand, the English still used the longbow, great carved sticks of yew wood that required years and years of practice and enormous muscle strength to use effectively.

Given the discrepancy in men and equipment, the outcome of the Battle of Agincourt was one of the most surprising in military history. Twenty-five thousand Frenchmen and allies, using crossbows no less, were soundly defeated by a force one-fifth their size utilizing the simple English longbow. There were many reasons for this, ranging from the blazing speed with which a longbow can be reloaded and shot, to superior English tactics and, most importantly, to rainy weather that muddied the fields so much that the heavy, armored French knights sank in the muck up to their hips.

For all these reasons, the French lost 15,000 knights and soldiers at Agincourt, while the English lost only 300 men. It was so lopsided and impressive a victory that Shakespeare immortalized the incident in his play *Henry V*. The French were so impressed and so enraged by the English longbow that they began a long-standing threat to summarily amputate the two fingers that hold the bowstring of any captured English archer.

That was quite a vicious threat, but the English were not easily cowed. They responded by waving their index and middle fingers in an insulting manner at the French from the top of their ramparts, as if to say, "My fingers? Here they are! Try and get them." To this day, this particular salute—two curved fingers raised in a *V*, held up to show the back of the hand and motioning up and down—remains in the English gesture lexicon as a terrific insult.

▷ 8 ◁

Pnewton's Petard

For tis the sport to have the engineer
Hoist with his own petard: and't shall go hard
But I will delve one yard below their mines,
And blow them at the moon.

—*Hamlet*, Shakespeare

"Hoist with his own petard" means that a person was dealt a blow by his own invention or machinations. When Shakespeare refers to the petard in the *Hamlet* quote above, he is speaking of a small, bell-shaped explosive charge. The military engineers of Prince Hamlet's era used such means to breach their enemy's fortified castle walls.

There was a big problem with fifteenth-century petards; namely, they were more dangerous to the maker than they were to the target. The petards were fairly crude devices and not usually made with any real care or precision. Therefore, they were prone to explode prematurely. When an unfortunate commando was blown sky-high by a premature petard blast, the enemy troops watching from the ramparts joked that another one had been "hoisted."

French armies first developed the petard concept. In fact, the word "petard" comes from the French word "peter," meaning "to break wind," or put less delicately but more accurately, "to fart."

In this chapter, we further develop the ideas introduced in the second chapter to build an air-powered potato launcher. The pneumatic-powered device in this chapter is called "Pnewton's petard," in honor of Sir Isaac Newton and his contributions to the physics behind this book, and because that propulsive force used here shares a similar theme with the word origin of petard. Again, we'll use PVC pipe and pipe-joining techniques to craft this device. Pnewton's petard consists of a pressure reservoir, a pressurization valve, a barrel, and a trigger valve. It can heave an Idaho russet halfway from Boise to Pocatello (well, almost) with surprising power, and with very little noise compared to the original (combustion-powered) potato cannon. This is stealth technology for backyard boomers.

Backyard boomers have made pneumatic cannons, big and small, for a long time. Some models built by hobbyists use an electronically activated valve called a "solenoid" to release air pressure into the cannon's barrel. The biggest air-powered cannons are so powerful that ski patrols and mountain rescue teams use them to break off huge, dangerous overhangs of snow from the sides of mountains, thereby reducing the risk of avalanches.

The model described here is simple to make, easy to operate, and—unlike some of the more exotic guns seen in the arsenals of other boomers—requires no electrical parts like a solenoid valve. The most difficult part about making Pnewton's petard is keeping all the connections clean and square so they are completely airtight and solid.

WORKING WITH PVC PIPE

Time for a quick refresher on working with PVC. Remember, PVC pipe is easy to use, but you must follow the instructions carefully. PVC pipe and the joining pieces that hold the pipes together are rated by the companies that make them to withstand pressures well over 100 pounds per square inch (PSI). However, in order for the PVC to hold pressures like this, you need to pay careful attention to the manufacturer's directions for joining the pipe. Read the PVC cement label carefully and follow all directions for joining pipe together. This includes priming the pipe, seating the pipe fully when you insert a pipe into a connector, and observing directions for cure times.

Take careful note: Although schedule-40 PVC is rated to withstand pressures of well over 100 PSI, you must limit the pressure inside the pressure reservoir to that shown in the directions. The solvent-welded parts and joint pieces may test safely to considerably less pressure than the pipe's pressure rating. Remember that safety is the most important concern and other factors such as air temperature and the age of your raw materials can affect the petard.

THE PIPE

As in the potato cannon project, this project uses polyvinyl chloride plastic pipe, usually called PVC. This project requires pressure-rated, schedule-40 PVC pipe and a couple of cast-iron pipe nipples.

THE CONNECTORS

We need more PVC connectors in addition to the couplings, bushings, and end caps discussed previously.

- ► Tee connectors—Tees are connectors that join three pipes together in the shape of capital *T*. If all the pipes are the same size, it is a straight tee. If one is smaller, it is a reducing tee.
- ► Pipe nipples—Pipe nipples are short pieces of male-threaded iron pipe. We'll use them to join the trigger valve to the rest of the petard.
- ► Male pipe and female end adapters—End adapters attach to the end of a smooth pipe and adapt it to either male or female pipe threads.

Building the Petard

Return to the hardware store where by now you have perhaps made friends with the plumbing clerk. Ask him or her to find the items in the following list. (You may have to go to an auto parts store for the replacement-tire air valve.)

All of the items shown in the materials list are quite common. If the store is out of any of the items you need, resist the temptation to "work around" the missing part by substituting two or more other parts that get you to about the same place. The fewer parts you use, the lower the risk of leaks. Besides that, the substituted parts may not work safely. If the store is out of a specified part, go find it at another store.

Materials

- ● Shaping file

- ● (1) 22-inch length of 1½-inch diameter schedule-40 PVC pipe

- ● (1) 1½-inch diameter male-threaded PVC pipe adapter

- ● PVC primer

- ● PVC cement

- ● Teflon pipe tape

- ● (1) 1½-inch diameter female-threaded PVC pipe adapter

- ● (2) 1½-inch diameter to ¾-inch threaded PVC reducing bushing

- ● Electric drill with various-sized drill bits (Drill sizes vary according to the diameter of the air valve and pressure gauge stem.)

- ● (1) rubber-coated, narrow-diameter, replacement-tire air valve

- ● (2) 3-inch-diameter PVC end caps

- ● (1) 0–60 PSIG (PSI gauge) with threaded bottom stem and nut

- ● Wrench

- ● (2) 9-inch-long pieces of 3-inch diameter Schedule-40 PVC pipe

- ● (1) 3-inch x 3-inch x 1½-inch PVC tee connector

- ● (2) ¾-inch diameter short iron pipe nipples

- ● (1) ¾-inch ball valve

- ● (1) 3-foot length of 1-inch diameter wooden dowel or broom handle

- Duct tape, optional
- Foot-stabilized air pump
- Bag of potatoes
- Earplugs and safety glasses

Place all of the materials and tools in front of you. Cutting, assembling, and filing will take about two hours. If you can enlist an assistant to help hold and file the pieces, the job will take less time. It is important that all connections are primed and solvent welded with PC cement throughout the entire diameter of the pipe so that the whole thing is airtight. Because the connections will be placed under pressure, the PVC-solvent weldments need to cure overnight.

Making the Barrel Assembly
1. Use the file to taper one end of the 22-inch-long, 1½-inch-diameter pipe section so it forms a sharp edge. The edge will cut the potato as it is rammed into the muzzle of the gun.
2. Attach the untapered side of the 22-inch-long pipe you just filed to the 1½-inch male pipe thread adapter according to the directions on the PVC cement can.
3. Attach the 1½-inch female adapter to the 1½-inch to ¾-inch reducing bushing according to the directions on the PVC cement can directions. Lightly screw the male adapter to the female adapter and put the whole barrel assembly aside to cure.

Making the Pressure Chamber
1. Drill a hole in the center of one 3-inch end cap just slightly smaller than the diameter of the tire valve. Replacement tire valves are coated with a layer of thick

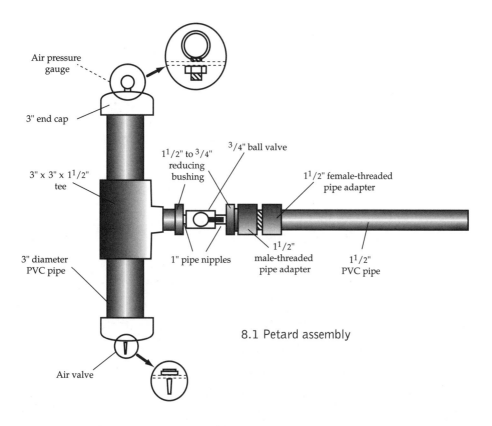

Air pressure gauge

3" end cap

$1^1/2$" to $^3/4$" reducing bushing

$^3/4$" ball valve

$1^1/2$" female-threaded pipe adapter

3" x 3" x $1^1/2$" tee

3" diameter PVC pipe

1" pipe nipples

$1^1/2$" male-threaded pipe adapter

$1^1/2$" PVC pipe

Air valve

8.1 Petard assembly

8.2 Component photo detail

rubber. You should be able to push the valve through the hole; it will snap into place and form a good air seal. (See diagram 8.1.)

2. Drill a hole in the center of the other 3-inch end cap for the pressure gauge. It should be slightly smaller than the diameter of the screw threads of the gauge. Next, screw the gauge into the end cap until it bottoms. Be sure to use the square brass fitting, which is usually present on the bottom of all gauges, for turning the gauge—*not* the face of the gauge. Secure the gauge into place with the brass nut that comes with most pressure gauges. Fasten the nut to the gauge threads securely using a wrench [see diagram 8.1].

3. Attach each end cap to one of the 3-inch diameter pipes by using the PVC cement.

4. Solvent weld the other end of the 3-inch pipes to the 3-inch openings in the tee connector using the PVC cement.

5. Take the remaining 1½-inch to ¾-inch reducing bushing and attach the 1½-inch end to the 1½-inch opening in the tee using PVC cement. This completes the pressure chamber assembly.

Final Assembly

1. Wrap the threads of the ¾-inch pipe nipples with Teflon pipe tape. Next, insert the nipples into each side of the ball valve as shown in diagram 8.1.

2. Now, screw in the ball valve–pipe nipple assembly to the 22-inch barrel.

3. Screw the whole barrel assembly into the pressure chamber assembly. Remember that all connections must be airtight.

4. Allow the entire assembly to cure overnight.

5. For additional safety, wrap the pressure reservoir longitudinally with three layers of duct tape. This provides a secondary layer of protection in the unlikely event of the solvent-welded end caps coming loose.

6. Close the valve and pressurize the reservoir to 10 PSI. Look and listen for leaks. Sprinkle soapy water on and around all joints. If you see bubbles, the joints do not have integrity; that is, they are not airtight. Unfortunately, there is no way of repairing leaky solvent-welded joints; you'll need to start over.

7. If the petard holds 10 PSI securely, pressurize the device to 30 PSI and retest all joints with the soapy water solution. If it passes this test, you're set to go. Aim the barrel in a safe direction and release the pressure by turning the valve.

8.3 Pnewton's petard

KEEPING SAFETY IN MIND

You must read and understand this section! Pnewton's petard has a lot of power, and it is important that you use it safely. It shoots potatoes with as much, if not more, force than the combustion-powered potato cannon. Therefore, the same safety rules are in effect.

1. As with all projects of this type, always use extreme care when aiming the device.
2. Make sure all end caps and other parts are securely attached with the proper cement using the proper procedures.
3. Check the cannon after every use for signs of wear or failure to make sure the barrel maintains its structural integrity. Replace any worn sections or parts immediately.
4. Do not overpressurize the petard. Thirty PSI is a reasonable maximum allowable pressure within the pressure reservoir. Although all of the individual components are rated higher, a high margin of safety should be maintained. The device is designed so that if an end cap did come loose, it would travel away from the user. Keep other people away from the area in front of the end caps and away from the front of the barrel.
5. This device can produce a moderately loud whooshing sound. Use ear protection and protective eyewear.
6. Clear the area in front of the cannon for 200 yards.
7. Clear the area around the cannon for at least 25 yards.
8. Yell a warning such as "Spuds away!" or "Fire in the hole!" before shooting just to make sure nobody walks into the field of fire.

OPERATING THE PETARD

The time to test the performance of the petard has finally come. By now you should have checked and rechecked your cannon and studied the safety procedures. You have a bag of potatoes, and your air pump is at hand. It's time to make another starch march and abuse some tubers.

1. Unscrew the barrel from the firing valve.
2. Using the wooden dowel or broom handle, ram a potato into the barrel, push it down to within five inches of the barrel's bottom, and reattach the barrel to the firing valve. Place the valve in the closed position.
3. Attach the foot-stabilized air pump to the tire valve. Keep a close eye on the pressure gauge and pump up the pressure within the reservoir to no more than 30 PSI. Keep your eyes and ears open for leaks. If you have a leak, you'll see bubbles and probably hear a hissing sound. The pressure gauge will indicate a leak by showing a steady decrease in pressure.
4. If everything seems OK, remove the air pump from the tire valve.
5. Taking careful note of the safety measures described above, aim the petard at a safe target. Keep in mind that the maximum range of such a device can only be determined through direct experimentation. Assume the spud can shoot 200 yards or more.
6. As soon as you're ready, shout the appropriate warning and then quickly snap the valve to the open position with your wrist. The powerful jet of air inside the pressure chamber will shoot the potato out of the barrel with a "whoosh!"

TIPS AND TROUBLESHOOTING

1. The petard has a narrower barrel than the potato cannon. So, you may get several potato projectiles from a single spud.
2. Wrap Teflon plumber's tape on the iron pipe nipple's screw threads to improve the seal.
3. Be sure to keep a close watch on the pressure gauge as you pressurize the petard with the air pump. Don't over-pressurize.

UNDERSTANDING PNEWTON'S PETARD

Interior ballistics, as you may recall from the earlier potato cannon chapter, is the study of the physics occurring inside the barrel of a shooting device. There are several obvious similarities between the combustion-powered potato cannon and Pnewton's petard, especially in terms of interior ballistics. Both use PVC tubing to control and direct the motion of the starchy vegetable plug, and both use expanding gas to propel the spud out of the barrel and toward a target.

In the spud gun or petard barrel, the pressurized gas accelerates the projectile due to the difference in pressure between the reservoir side of the device and the nonpressurized barrel side. If you remember the Newton's Laws discussion from the second chapter, we outlined how Newton's second law explains that the spud's acceleration within the barrel is equal to the force on the potato divided by its mass. Physicists express the second law as $F = mA$ or $A = F/m$, where F stands for force, m for mass, and A for acceleration. This equation shows us that the bigger the force, the more the potato accelerates down the barrel. There are opposing forces working inside the barrel to slow the potato, such as the frictional force between potato and barrel, but these are small compared to the pressure push from the air reservoir.

When the potato leaves the barrel, everything changes. There is no longer a high-pressure push, courtesy of the expanding gas inside the barrel. Once the potato leaves the barrel, Newton's first law takes center stage—a potato in motion stays in motion until acted upon by an external force. Of course, the spud won't fly forever. In this case, the potato will fly in a straight line, but it is continually acted upon by two other forces to stop it—gravity and air drag. Gravity pulls the tuber down to the ground and, simultaneously, the tuber has to push the air molecules out of its way as it flies. This is called air drag, and it works counter to forward motion.

Some people incorrectly think that the potato gathers speed as it flies through the air. They imagine the potato gathering force and momentum as it flies toward the target. Nope; not true. The fastest the potato ever flies is at the split second it leaves the barrel. After this, the air friction slows it down as it travels. (However, if you shoot it straight down, that's another story.) In level flight, the muzzle velocity at exit is as good as it gets.

▶ IN THE SPOTLIGHT ◀

SECRET WEAPONS

Pnewton's petard is a great leap forward in the field of clandestine potato hurling. While it's not really a secret, it did require a fair amount of research and development time, just like far more complicated engineering and science projects.

Most people have heard of or read about the Manhattan Project, the top-secret weapons project that led to the development of the atomic bomb. Every era and every war has its own secret weapons. Some were world-changing devices that shaped the time lines of history. Others seemed like good ideas to their creators before falling into the dustheap of historical military trivia. Beginning with the rock-slinging engines of Archimedes and continuing through the recent "super gun" of Saddam Hussein, the list of secret weapons is long and intriguing.

One very important nineteenth-century military advance involved improvements in rocketry based on an English engineer's fascination with Indian military ingenuity. Around 1804, a British inventor named William Congrave heard stories from returning English soldiers about rockets in the Indian Raj.

There, Hyder Ali, prince of India's Mysore kingdom, had developed military rockets with an important improvement: He used metal cylinders to contain the combustion process and a long pole to move back the rocket's center of gravity. The metal propellant container was far superior to the paper tubes used previously. The lower center of gravity made the rocket fly straighter. Even with these advances, a single rocket, by itself, was not particularly accurate or militarily effective, but when fired in massive numbers against troops, they were incredibly frightening. Hyder Ali and later his son, Tippu Sultan, used these rockets with considerable success against British colonial troops.

The news of the successful use of rockets spread throughout Europe. In England, Congrave began to experiment with different rocket designs. His experiments led to standardized construction techniques and accurate and reproducible black-powder formulations. Congrave rockets, as they came to be called, eventually evolved to the point where they could accurately deliver either an explosive payload or an incendiary warhead. The Smithsonian National Air and Space Museum collection displays Congrave rockets. They look like 10-foot-long wooden matchsticks with black metal heads. The English army found them devastatingly effective when launched en masse against military targets.

Congrave's rockets were employed in a successful naval bombardment of the French coastal city of Boulogne in 1806. The next year, a very large military siege—using thousands of Congrave rockets—burned Copenhagen, Denmark, to the ground.

In the United States, the British Redcoats made extensive use of Congrave rockets during the War of 1812. The English, with their considerable industrial might, were able to produce huge numbers of rockets. The British used the rockets most effectively in two key battles. In 1814, during the Battle of Bladensburg, the British rockets played a key role in the defeat of American troops defending Washington, D.C. In September of the same year, the British forces attempted to capture Fort McHenry, which overlooks Baltimore Harbor. The British admiralty uncovered its latest secret weapon, a specially designed ship named the *Erebus*, to launch the Congraves. Over the two days, beginning on September 13, the American defenders of Fort McHenry hunkered down and withstood rocket and artillery barrages unleashed by the *Erebus* and four other "bomb ships." Eventually, they returned fire from the fort's complement of cannons and forced the British men-of-war to withdraw.

The pitched battle inspired Francis Scott Key to write "The Star Spangled Banner," which was later adopted as the

United States national anthem. "The rockets' red glare" continues to remind us of Congrave's rockets.

One hundred years later, military leaders on both sides were looking for a way to break World War I's trench warfare stalemate on the Western Front. Both sides wanted something special, a great military leap forward like the Congrave, to get them out of the Great War's horror of mud, rats, and trench foot. The German general staff decided a terror weapon, to be used against the city of Paris, might be the way to turn the tide of battle. In response to this idea, the German arms-makers built the largest artillery device in history.

In early 1918, the largest cannons extant could hurl a massive explosive shell about 23 miles, or the approximate distance of an Olympic marathon. It takes a long-distance runner about two and a half hours to cover such a distance; the big guns of WWI could hurl a two-hundred-kilogram-high explosive shell the same distance in 80 seconds.

The German generals thought that if they could find a way to bombard Paris, the terror that would ensue could change the entire outcome of the war. The French government, they reasoned, would be heavily pressured by a suddenly vulnerable civilian populace and would be forced to negotiate for peace on terms favorable to Germany. But the German artillery emplacements were much farther than 23 miles from Paris. Hence, the need for a super long-range gun. The munitions makers set to work on designing this super gun, which the Allied forces called the "Paris Gun."

The Germans called it the "Kaiserwilhelmgeschutz," which translates to "Kaiser Wilhelm's gun." It was designed and built in an amazingly short time by a crack team of Germany's most experienced gun designers. After little testing, it was used to shell Parisian suburbs from March 1918 until August

1918. The Germans positioned the guns on movable railway cars located in the dense forests of Cercy, about 70 miles from the city.

The Paris Gun looked great on blueprints, but it never worked nearly as well as the generals had hoped. There were seven barrels made but only two mountings; so, effectively, only two guns were ever in use at any time. Also, the massive guns developed such high internal pressures and temperatures that they wore out very quickly. Most importantly, at such great distances, the accuracy of the Paris Gun was terrible. It could never come close to hitting a specific target and was of value only as a way to dishearten and frighten the French civilians.

The shells weighed nearly 220 pounds and had a range of 90 miles. The Paris Gun could have had a major impact on the course of the war if it had been more accurate and if the Germans could have built more of them. The Paris Gun did meet with some small success—it did terrify France and kill some Parisian civilians—but in the end it had little effect on the war's outcome. The guns were withdrawn and dismantled in the face of the Allied advances in August 1918. One spare mounting was captured by American troops near Chateau-Thierry, but no gun was ever found by the Allies during or after the war.

▷ 9 ◁

The Dry Cleaner Bag Balloon

During the American Civil War, Count Ferdinand von Zeppelin was an official military observer of the Union army for the German government. He was not an active participant in the war, but he was an excellent observer. He recorded everyone and everything that could possibly be of use to his Prussian superiors back in Berlin. One thing he especially noted was the use of manned balloons by the warring armies.

The United States first used balloons for military purposes during the Civil War. While balloonists for both the North and South carried out important military missions early in the war, the use of balloons ceased about halfway through the conflict. Union military commanders didn't really understand or appreciate the advantages the balloon-ists gave their artillery spotters or intelligence corps and sub-sequently shut down the Army Balloon Corps in 1863. The South may have seen more value in the balloons but simply couldn't afford to build them.

After they lost their army jobs, many of the former bal-loonists took their airships and became barnstormers. The barnstorming balloonists went from town to town putting on air shows for entertainment-starved townsfolk. They per-formed such feats as dropping parachuted animals from bal-loons, swinging from balloon-mounted trapezes, and heaving fireworks over the side of the craft for enthusiastic crowds below. They also offered balloon rides for a price.

Zeppelin remembered what he saw the balloon aircraft do in the Civil War battlefields of 1862 and sought out the barnstormers for more information. Zeppelin took his first balloon flight lesson in St. Paul, Minnesota, from an ex-Union soldier named John Steiner. Steiner sent him up on a solo balloon flight at the end of a 700-foot tether rope. The young officer wrote a report of the experience to his superi-ors back in Germany. In the report, Zeppelin's enthusiasm is subdued, and it appears to be only a simple reporting of the military potential of observation balloons. But, deep inside, the experience changed Zeppelin forever. He found balloons fascinating and exciting, and studying them would become a priority.

Soon after the flight, he returned to Germany and spent the next 20 years in various military roles before he returned to work on his idea of commercializing travel via lighter-than-air craft. In 1887, he published a detailed plan for a European network of airships to take passengers from one city to another. Zeppelin was a driven entrepreneur and soon began building rigid airships in his own factory. By 1908 his dirigibles were flying on commercial routes throughout Germany, transporting mail, cargo, and people.

For the next 30 years, the spectacular Zeppelin airships dazzled everyone. Zeppelin air travel was the most glamorous and prestigious form of personal transport until the lighter-than-air transportation industry collapsed after the explosion of the hydrogen-filled Zeppelin *Hindenburg* in 1937.

Building the "Jellyfish of the Sky"

Like Count von Zeppelin's rigid airship, the dry cleaner bag balloon, or flying jellyfish, demonstrates the same principles of buoyancy first explored by the ancient Greek Archimedes. The device is made from a small aluminum foil cupcake tin, magnet wire, jellied alcohol, and a *very* light plastic bag. The thin bags that dry cleaners use to protect freshly cleaned suits are large enough and light enough to work well. Plastic trash bags and leaf bags are too heavy; the bag must be extremely lightweight.

The foil cupcake tin holds a small quantity of solid alcohol fuel. The ignited fuel heats the air inside the bag enough to make it buoyant. As the inside air temperature increases, the bag slowly inflates, hovers, and then gracefully rises to an impressive altitude. The clear bag hovering and floating in the air often reminds people of an airborne jellyfish.

The 1892 *All-American Boy's Handy Book* contains a lengthy section on how boys made hot air balloons from precisely cut sheets of paper that were glued together, fitted with a heat source, and flown. Making the precise cuts and folds on the paper takes a great deal of time and demands painstaking accuracy. This project's updated design uses "modern" technology—a plastic bag—and is easier and faster. However, this balloon design is fairly crude; it flies well but only in the right weather conditions.

KEEPING SAFETY IN MIND

The safety tips that apply to the Cincinnati fire kite apply even more stringently to the flying jellyfish. Pick the right time and place to launch your balloon. If the weather conditions are right, the balloon floats up in the air, hovers, and then descends gently. There are government regulations that prohibit flying balloons in an unsafe manner, so be sure to observe the following safety rules.

1. Fly the balloon only when the air is calm. Do not fly the balloon when breezes aloft could send it into hazardous areas.
2. Pick the right place to launch the balloon. The right place is a large open area where there is no possibility that the balloon will contact any other objects.
3. Do not fly balloons near areas where aircraft fly.
4. In general, use common sense and do not fly the balloon in locations where it could contact dry areas, houses, trees, buildings, or anything else that would cause damage or hurt your relationship with your neighbors.

Materials

- (1) aluminum foil cupcake tin, 2-inch to 2½-inch diameter
- Awl or ice pick
- (4) 12-inch lengths of magnet wire (26-gauge wire)
- Polyethylene dry-cleaner bags
- (1) full spool of extra-strong nylon thread
- Jellied alcohol (sold under brand names such as Canned Heat or Sterno)

1. The overall weight of the entire balloon assembly—plastic bag, wires, fuel, and aluminum cup—should be less than 40 grams. If required, you can reduce the weight of the cupcake tin by removing the upper edge and lip with a tin snip or heavy shears. If you do this, fold the cut edge over to prevent nicks or cuts. Refer to diagram 9.1.
2. Poke four small holes, one every 90 degrees, in the upper edge of the cupcake tin. Insert the 12-inch magnet wires and secure by folding and knotting the wire.
3. Pinch together the polyethylene bag edges into a small bunch. Loop the end of the magnet wire a few times around the bunched edges and tie off. This completes the balloon assembly process.
4. Take the balloon outside. It is now ready to launch.

LAUNCH, LEVITATION, AND LIFTOFF

1. Place 2½ grams of jellied alcohol in the bottom of the cupcake tin. Distribute the jellied alcohol around the cupcake tin so there is a reasonable amount on the surface

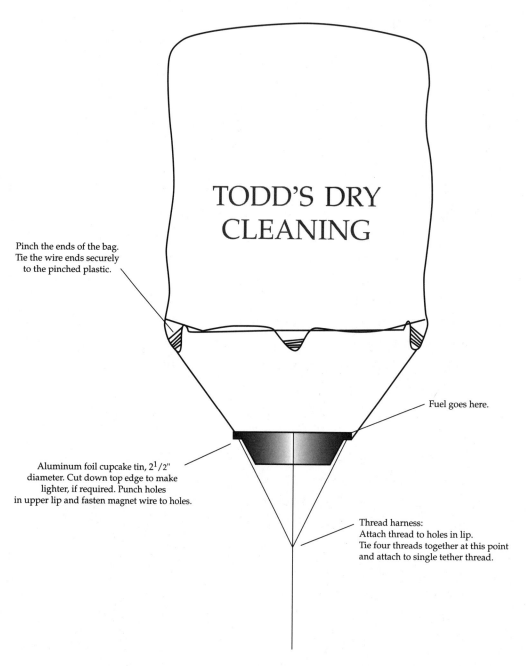

TODD'S DRY
CLEANING

Pinch the ends of the bag.
Tie the wire ends securely
to the pinched plastic.

Fuel goes here.

Aluminum foil cupcake tin, $2^1/2"$
diameter. Cut down top edge to make
lighter, if required. Punch holes
in upper lip and fasten magnet wire to holes.

Thread harness:
Attach thread to holes in lip.
Tie four threads together at this point
and attach to single tether thread.

9.1 Dry cleaner bag balloon assembly

area. (A "reasonable amount" is about the size of a quarter, but experiment to see what works best.)

2. Make a harness for the cupcake tin using the thread as shown in diagram 9.1. Tie the thread harness to the holes in the cupcake tin that you used to fasten the magnet wire. The thread is only used to make the balloon easier to retrieve. However, the thread may snap, break, burn, etc. Do not depend solely on the thread harness to prevent the balloon from floating into dangerous areas. Make sure you pick a safe area for the launch.

3. Place the aluminum cupcake tin on the ground. Have two helpers hold the bag open over the tin plate while you light the jellied alcohol. As the alcohol burns, hot air will slowly fill the plastic bag and cause it to expand.

4. In a short time, the hot air will cause the balloon to become buoyant. It will reach a state of neutral buoyancy; that is, the buoyant force pushing it up will just be offset by the force of gravity. The balloon should hover, almost magically, in midair.

5. As the air inside the bag continues to heat, the buoyant force overcomes gravity and the balloon rises. Play out the thread steadily as it rises.

6. In a few minutes, the fuel will run out. The bag will cool down and float gently back to earth.

TIPS AND TROUBLESHOOTING

The flying jellyfish can float up to 40 feet high. Its ultimate height depends on the amount of fuel, the size of the bag, the weight of the cupcake tin and thread, and atmospheric conditions. As with the Cincinnati fire kite, atmospheric conditions (for example, temperature, humidity) play a big role in determining how fast and how high the device rises.

BALLOON PHYSICS

What is the difference between engineers and scientists? Here is an old story that somewhat explains:

> Once upon a time, a mathematician, a physicist, and an engineer were all given a red rubber ball and told to find the volume. The mathematician carefully measured the diameter and evaluated a triple integral. The physicist filled a beaker with water, put the ball in the water, and measured the total displacement. The engineer looked up the model and serial number in his red-rubber-ball table.

The buoyancy theorem, better known as the Archimedes Principle, states that the upward force on an immersed object is equal to the weight of the displaced fluid. What exactly does this mean? Well, applying the Archimedes Principle to our balloon, it means that hot air balloons rise because the balloon and the volume of heated air inside it together weigh less than the same volume of unheated air. This difference is referred to as "lift." Lift tells us how much payload our balloon could carry.

Let's calculate the maximum amount of lift the balloon produces. In order to do this, we need to first weigh the cupcake tin, magnet wire, fuel, and plastic bag. For this example, let's assume the total weight is 1½ ounces. Next, we need to find the volume of our bag. To simplify the calculation, we'll use round numbers only and assume the bag can be represented as a cylindrical volume, say 3 feet long and 2 feet in diameter. The volume of any cylinder is figured by finding the area of the base and multiplying that by the cylinder height. For our bag:

Volume of Bag = $\pi \times 1$ foot^2 $\times$ 2 feet = 6 cubic feet.

Next, we need to measure the air temperature inside and outside the balloon. For this example, assume that the outdoor air temperature is 60°F and a thermometer reading tells us that the fuel heats the inside of the balloon up to approximately 160°F.

Now we have enough information to calculate the upward force vector, F_{up}:

F_{up} = (the balloon volume) x (the difference in density between the heated and unheated air).

Physicists would calculate the density of air at different temperatures by using an equation called the "ideal gas law." Engineers, being very practical people, would normally forgo such arduous calculations and just look up the values in a handy reference table of air densities. According to the table, the density of air at 60 degrees and at regular atmospheric pressure is .076 pounds per cubic foot. At 160 degrees, it is in the neighborhood of .061 pounds per cubic foot. Therefore, the lifting force is: 6 cubic feet x (.076 − 061) = .08 pounds up.

The force down is the weight of the balloon, or .06 pounds down.

The net lifting force is .02 pounds. Because the net force is upward, the balloon will rise. If we added additional payload weight of .02 pounds, it would hover in midair. If the balloon and its payload exceeded the .08 pounds-force lifting force, the balloon would not rise.

▷ 10 ◁

The Carbide Cannon

What was it like to be an underground miner in the eighteenth or nineteenth century? Certainly, it was dirty, it was loud, and it was dangerous, but the most pervasive unpleasant aspect of mining was the dark. Candles and oil lamps provided what dim illumination there was. Slinging a ton and a half of rock into a narrow, steel-wheeled truck in the lighting equivalent of an eight-watt bulb was the grim daily reality of life underground.

Lighting in the mines was crucial to miners, since better lighting meant better safety conditions and higher productivity. A great deal of effort was spent to improve lighting conditions.

Mine lighting followed an evolutionary path. Miners in the 1700s used a lamp with a wick laid in grease or animal fat. These lamps were dim, smoky, and hard to use. Early miners had to keep a very sizable supply of matches close at hand because even a slight breeze left the miner stumbling his way through the darkness. Eventually, lamp manufacturers took the logical step of outfitting mining lamps with transparent covers to protect the flame from wind.

Oil lamps replaced solid fat-burning lighting, as they were cheaper to use than the tallow candles and also easier to balance and carry in the mines. They differed in shape and size, but they all shared the same basic principle of operation. A small conical reservoir held the fuel, and on top of it a hinged cap snapped on to seal the top. There was a long spout that extended up and outward from one side on the reservoir. On the side opposite the spout, a wire hook connected to the reservoir and latched onto the miner's leather or cloth protective helmet. It looked like a small teapot with a brush hanging out the spout. The wick brought the fuel from the font to the tip. This was considerably more convenient and efficient than a simple candleholder. But, shortly after oil lamps, the carbide lamp came into general use. It provided miners with what they really needed—bright, dependable, and safe mine lighting.

In May 1892, Major James T. Morehead and Thomas L. Willson were struggling to make aluminum in an electric furnace of their own unique design. Their efforts to that point had not been economically successful, so they tried a different tack. They focused their attention on producing metallic calcium. The metallic calcium, if produced in sufficient quantity, could in turn be used to produce aluminum inexpensively. Unfortunately, Willson and Morehead found that it was even more difficult to economically produce metallic calcium than it was to produce aluminum.

However, "Carbide" Willson, as he came to be called, was lucky. He poured coal tar and lime into a furnace, expecting to produce metallic calcium. The end result of Carbide's experiment was a gray, brittle substance that sizzled up with gas when it was placed into water. The gas produced was flammable, and Willson named it "acetylene."

The gray substance turned out to be the chemical compound called calcium carbide. Willson and Morehead knew a winner when they saw it and quickly changed their business's direction. They sold rights to the process to eight companies that wanted to produce calcium carbide in great quantities to make acetylene for lighting and industrial applications. Seven of these companies couldn't make a profit. The last company built a chemical plant in northern Michigan and was indeed extremely successful. The company, Union Carbide, built a chemical empire out of calcium carbide and acetylene and went on to become one of the largest companies in the world.

In 1890, licensees of Willson and Morehouse offered the first calcium carbide lamp, and soon after that the lamp was adapted for underground mining. Carbide lamps were a great improvement in mine lighting technology. The first carbide mining lamps burned for approximately four hours with a one-inch silver-white flame. Not a tremendous amount of light, but much better than attaching an oil lamp to the top of your head. Unlike oil-wicked devices, carbide lamps could be fashioned into headlamps and could direct the light where the miner needed it most.

In the 1920s and 1930s, electric lights replaced fuel-burning lamps in most mining operations. However, carbide lamps continue to be manufactured and sold to spelunkers, who prize them for their light weight, dependability, and historical value.

A carbide lamp fits on the front of a safety helmet, providing intense illumination just in front of the spelunker's face. The carbide lamp works like this: A small, gray, rocky chip of calcium carbide is placed in a small metal container, called the gas generator. A regulated amount of water is constantly dripped on top of the carbide chunks inside the generator. When the water and calcium carbide mix, a chemical reaction takes place and forms flammable acetylene gas. By collecting the gas and directing it through a pinhole-sized nozzle, it can be burned, and when the gas burns, it does so with an intense, cutting torch-like white light. The good thing about the carbide-water reaction is that it is relatively easy to regulate and is safe to wear around your head in a lamp.

The carbide cannon is a direct descendant of the carbide headlamp. This device is yet another easy-to-build PVC pipe project. Instead of shooting potatoes or tennis balls, the carbide cannon's appeal lies in its ability to make an incredibly loud bang and bright flash of light. The flash-bang of a good carbide cannon is both exciting and inspiring. With a little imagination, it isn't too hard to see in it what Napoleon saw in his best artillery and what Admiral William "Bull" Halsey saw in the battleship *Missouri*'s 16-inch guns.

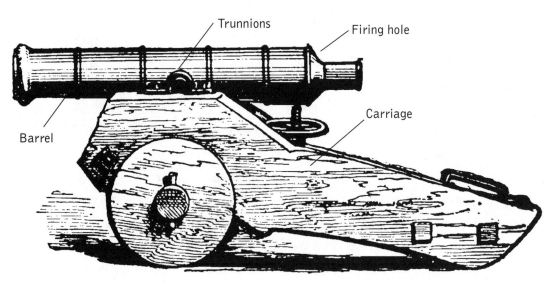

Trunnions

Firing hole

Carriage

Barrel

10.1 Carbide cannon

Building the Carbide Cannon

This project also makes extensive use of PVC pipe and pipe fittings.

Cutting, gluing, and assembling the carbide cannon will take about two hours. This is a terrific project for parents and their older children, and it is arguably a safe alternative to firecrackers for Fourth of July entertainment. It is always more fun to work with a helper, and the extra pair of hands will speed up the assembly and increase the cannon's accuracy. Remember, just as with the potato gun, you'll need to let the PVC cement cure overnight before using the cannon.

This is important: Assemble the cannon twice. First, dry fit (with no cement) all the parts to make sure you understand how everything fits together. Then, once you've got it all figured out, do it a second time, making nice, neat, tight joints using the PVC cement. Pay close attention to the assembly drawings provided.

Materials

Note: All pipes and fittings are schedule-40 PVC.

- Hacksaw

- Twist drill

- Long-handled fireplace match or piezoelectric lighter

- (6) ¾-inch diameter elbow fittings

- (2) ¾-inch diameter tee fittings

- (1) 2-inch to 1½-inch flush reducing bushing

- (1) 1$\frac{1}{2}$-inch to 1-inch flush reducing bushing
- (2) $\frac{3}{4}$-inch inside diameter coupling (with one end smooth and one end female-threaded)
- (2) $\frac{3}{4}$-inch diameter male-threaded adapter couplings (with one end 1-inch outside diameter and one end $\frac{3}{4}$-inch male threaded)
- (1) 2-inch diameter cross
- (1) 1-inch diameter cross
- (3) 2-inch diameter threaded end caps
- (3) 2-inch diameter threaded adapters
- (1) can PVC cement
- (1) can PVC primer
- (2) 5-inch lengths of $\frac{3}{4}$-inch PVC pipe (uprights)
- (2) 2-inch lengths of $\frac{3}{4}$-inch PVC pipe (trunnions)
- (2) 12-inch lengths of $\frac{3}{4}$-inch PVC pipe (crosspieces)
- (2) 4$\frac{1}{2}$-inch lengths of $\frac{3}{4}$-inch PVC pipe (short longerons)
- (2) 9$\frac{3}{4}$-inch lengths of $\frac{3}{4}$-inch PVC pipe (long longerons)
- (1) 15$\frac{1}{2}$-inch length of 1-inch PVC pipe (barrel)
- (1) 5-inch length of $\frac{1}{2}$-inch PVC pipe (loader)
- Earplugs
- Calcium carbide

Note: Depending on where you live, you may be able to purchase carbide chips at stores that cater to underground cave explorers. But it may be more convenient to purchase powdered calcium carbide under the trade name Bangsite at

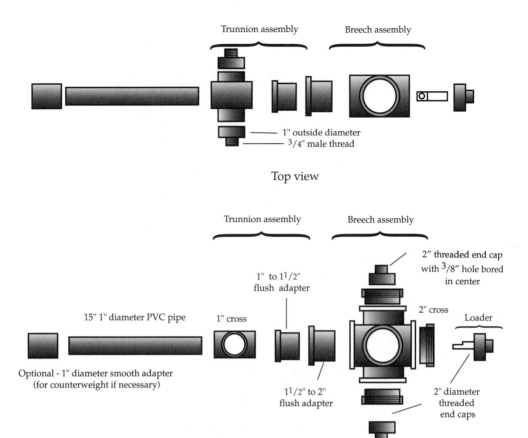

Top view

Side view

10.2 Cannon barrel assembly

your local hobby store or via mail-order. Mail-order Bangsite is available from the following company:

The Conestoga Company, Inc.
P.O. Box 405
Bethlehem, PA 18016-0405
1-800-987-BANG (2264)

It is likely that other suppliers of Bangsite can be found on the Internet by going to a major search engine and searching under Bangsite.

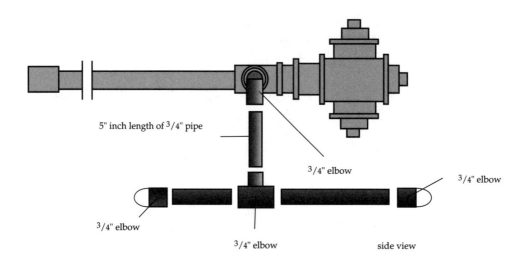

5" inch length of ³/4" pipe

³/4" elbow

³/4" elbow

³/4" elbow

³/4" elbow

side view

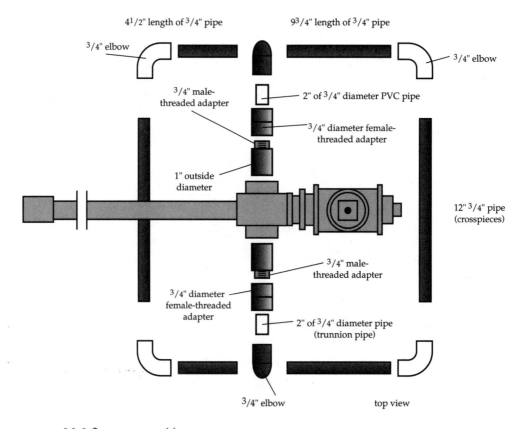

4¹/2" length of ³/4" pipe

9³/4" length of ³/4" pipe

³/4" elbow

³/4" elbow

³/4" male-threaded adapter

2" of ³/4" diameter PVC pipe

³/4" diameter female-threaded adapter

1" outside diameter

12" ³/4" pipe (crosspieces)

³/4" male-threaded adapter

³/4" diameter female-threaded adapter

2" of ³/4" diameter pipe (trunnion pipe)

³/4" elbow

top view

10.3 Cannon assembly

10.4 Carbide cannon

PREPARING THE MATERIALS

1. Lay out all the PVC pipe fittings—the tees, the crosses, the elbows, the threaded couplings, and so forth neatly in front of you.
2. Cut the PVC pipe pieces to the lengths specified in the materials list. Be as accurate as possible with your saw.
3. Use special-purpose PVC primer and PVC cement and follow all label directions.

Making the Cannon Assembly

Begin by making the cannon assembly. The cannon assembly consists of two subassemblies: the breech subassembly and the trunnion subassembly. (The breech is the back part of the cannon and the trunnions are the supports that hold the barrel and allow it to change elevation.)

139

Making the Trunnion Subassembly

1. Place and cement the adapters that connect the 1-inch smooth pipe to the ¾-inch threaded pipe on opposite sides of the 1-inch cross. (These pipe threads will later engage the female pipe threads attached to the carriage. This forms the actual trunnions.)

2. Place and cement the 1-inch to 1½-inch flush reducing bushing into one of the open sides of the 1-inch cross. This completes the trunnion subassembly.

3. Cement the 1-inch PVC pipe (the barrel) into the remaining opening of the 1-inch cross. Make sure the cannon barrel is fully seated into the opening.

Making the Breech Subassembly

1. Cement the 2-inch threaded adapters into three of the four openings of the 2-inch cross.

2. There are three threaded end caps that screw onto the threaded adapters you cemented into place in the previous step. Each end cap has a specific purpose. The bottom end cap will be filled with water and is the container in which the carbide and water react to form acetylene. The back end cap holds the carbide loader. (The instructions for making the loader assembly are in a later step.) The top end cap contains the firing hole.

3. Drill a ⅜-inch hole through the top threaded end cap as shown in the drawings. This is the firing hole.

4. Cement the 2-inch to 1½-inch reducing bushing into the remaining opening on the 2-inch cross.

5. Attach the breech to the trunnion by joining the 1½-inch female end of the reducing bushing projecting from the 2-inch cross to the 1½-inch male end projecting from the 1-inch cross. When you insert both assemblies into place, the crosses must be cemented such that the plane of one

cross lines up at 90 degrees to the plane of the other
cross. Refer to diagram 10.5 to see how the 2-inch cross
and the 1-inch cross align.

6. Allow the cannon assembly to dry.

Making the Carriage

1. Start by making the two upright supports. Cement the
   two 5-inch long, ¾-inch pipes into the middle opening of
   the ¾-inch tees. To do this, insert the pipe exactly ¾-inch
   deep into the run of the tee. While keeping the opening of
   the elbow at a right angle to the tee, cement the pipe onto
   the elbow. See the upper half of diagram 10.3 for details.

2. Next, make the pieces that hold the cannon that are parallel to the barrel. These are sometimes called
   "longerons." Cement the 4½-inch (short longeron) and
   the 9½-inch (long longeron) pieces of ¾-inch PVC onto
   the remaining open sides of the ¾-inch tees.

3. Cement a ¾-inch elbow on each remaining longeron end.
   The elbows must be carefully aligned as shown in diagram 10.3. The elbows must be aligned so that the openings are all facing right on the left longeron and just the
   opposite for the right longeron.

4. Place the 2½-inch long, ¾-inch diameter PVC trunnion
   pipes into the opening of the elbow at the top of the upright
   supports exactly ¾-inch deep into the elbow described in
   step 1 above. Cement into place. Cement the female
   threaded adapters onto the exposed ¾-inch PVC pipe
   stubs, again making the PVC pipe seat exactly ¾-inch deep
   into the socket end of the female threaded adapter.

Final Assembly

Attach the left and right uprights to the cannon assembly by
screwing together the male and female ends of the adapter

parts of the trunnions. Check that the trunnions are fairly tight but still turn smoothly.

1. Make two crosspieces. The crosspieces are a nominal 12 inches long. However, depending on how the male and female trunnions were built, the length of the crosspieces may vary. So, measure the total length of the trunnion assembly, consisting of the PVC trunnion pipes that are inserted into the joined male and female adapters, which in turn are inserted into the 1-inch cross. Use this as the length for the crosspieces. See the diagram for more details. Cut the crosspieces to size and complete the cannon assembly by cementing the crosspieces into place.

2. Let all joints cure overnight.

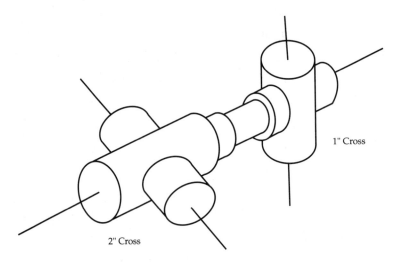

1" Cross

2" Cross

10.5 Crosspiece assembly

Making the Carbide Loader

1. Refer to the carbide cannon assembly diagrams. Cut a $4\frac{3}{4}$-inch length of $\frac{1}{2}$-inch diameter PVC pipe as shown in the diagram. You will cut away the top half of the first inch of pipe.

2. With a ¼-inch twist drill, drill a cup or bowl about ⅛-inch deep in the bottom of the cutaway area as shown in the diagram. This divot or cup is your measure for granulated calcium carbide. Be very careful as you drill; it is important that you do not drill all the way through the plastic pipe.

3. Insert the noncut end of the ½-inch PVC loader piece into the square depression on one of the 2-inch threaded caps. Cement it securely into place using PVC primer and cement.

4. If everything was done correctly, the carbide loader will dump the Bangsite into the threaded cap on the bottom of the barrel assembly (which will be filled with water) when the threaded cap is screwed securely into place. See the diagram to see how all of this fits together.

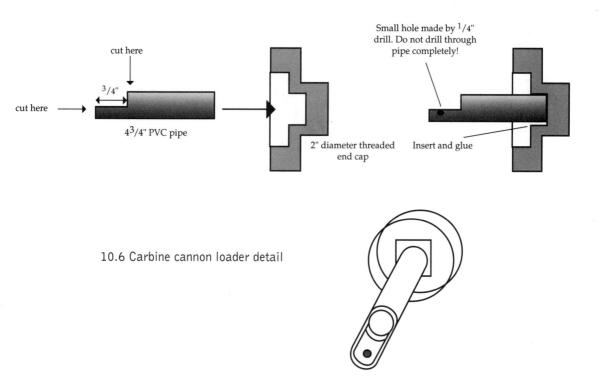

cut here

³/4"

cut here

4³/4" PVC pipe

2" diameter threaded end cap

Small hole made by ¹/4" drill. Do not drill through pipe completely!

Insert and glue

10.6 Carbine cannon loader detail

FIRING THE CARBIDE CANNON

Remember, always think carefully about what you're doing! You are responsible for your own safety. Read and understand this section before using the carbide cannon. Just like the potato cannon, this wonderful device you just made with your own two hands deserves respect. Read and follow these rules to ensure safety.

1. The acetylene gas produces considerable energy and should be treated with caution and respect. Place just enough Bangsite in the loader to fill the depression made in the previous step with the ¼-inch twist drill. The correct amount of Bangsite is roughly equal to ¹⁄₁₆ of a teaspoon. If you use less, the gun's performance will suffer. If you use too much, the concentration of gas will be too high and you'll have poor performance, waste fuel, and you'll expose yourself to needless risk.
2. Load the carbide cannon by screwing the loader into place. As you screw the threaded cap into the threaded adapter, you'll dump the Bangsite into the water-filled reservoir. Immediately, you will hear a fizzing sound. This is the acetylene forming in the combustion chamber.
3. The cannon is loud! Warn everyone that you're going to fire it, wear hearing protection, and keep the cannon away from other people.
4. Take a long-handled fireplace match or long-handled piezoelectric lighter and apply a flame to the touch hole. The cannon will fire with a loud, sharp report.

KEEPING SAFETY IN MIND

1. Never shoot the cannon at anybody or anything.
2. Don't allow the powdered calcium carbide to touch anything wet, such as your skin or eyes.
3. Don't eat or taste the calcium carbide.
4. This cannon is for making noise, not for shooting projectiles. Don't put projectiles into the gun barrel.
5. Never look down into the barrel of a loaded cannon.
6. Use caution when emptying the water chamber. The water will be white and cloudy from lime residue. The lime could stain clothes, furniture, etc.

WHERE DOES THE BANG COME FROM?

The interesting thing about calcium carbide is that it reacts with water to produce acetylene gas and lime. Lime is the white chalky residue left in the water at the bottom of the cannon. When you ignite the acetylene, a second reaction occurs. The flame starts an exothermic chemical reaction resulting in the release of energy, which your hear as a big bang, as well as the by-products carbon dioxide and water.

▷ 11 ◁

The Ballistic Pendulum

HOW FAST DOES IT GO?

You've probably noticed that many of the projects in this book involve a device that fires, shoots, hurls, or otherwise propels an object at a reasonably high velocity. After the novelty of the shooting experience wears off, inquiring boomers may want to experiment with the ballistic device called a "ballistic pendulum." With the pendulum, we can dig deeper into the mechanics of motion that govern all of the experiments listed in this book.

If you truly understand the mechanics of motion, you can answer all of the "how" questions—how far will it fly?, how high will it go?, how long will it travel?, and, *most intriguingly,* how fast does it go? Does the potato cannon shoot a spud faster than Randy Johnson throws a fastball? Faster than Tim Wakefield throws his knuckleball? Which is faster, a tennis ball shot from a mortar, or Boris Becker's crosscourt backhand? Does a walnut hurled from the tabletop catapult fly faster than the airspeed velocity of an unladen sparrow (African or European)?

The "how fast does it go?" question can be answered by using a ballistic pendulum. A ballistic pendulum uses two well-known Newtonian ideas—the principle of conservation of momentum and the principle of the conservation of energy—to allow you to easily determine the velocity of almost any projectile. Ballistic pendulums measure the velocity of bullets, cannonballs, catapulted boulders, or just about any nonpowered flying item.

Building the Ballistic Pendulum

The ballistic pendulum consists of a body of known mass suspended from some light wires, as shown in diagram 11.1. The ballistic pendulum allows you to measure three easily found parameters: the mass of the projectile, the mass of the pendulum, and the height the pendulum rises after being struck by the projectile. From them, you can calculate a fourth, much less easily measured parameter—the speed of the rapidly moving projectile.

Materials

- Scissors
- Ruler
- (1) cardboard box approximately 12 inches x 12 inches x 18 inches
- Duct tape
- Newspapers and weights to make the cardboard box weigh exactly 64 ounces (4 lbs)
- (4) eye hooks
- (4) 3-foot-long strands of nylon string
- (1) 4-foot x 4-foot piece of plywood or sturdy cardboard
- (1) felt tip marker
- (1) 4-foot x 4-foot sheet of white paper

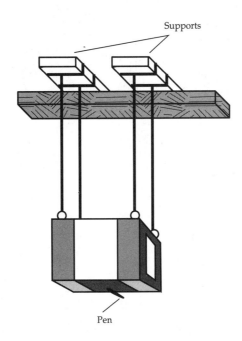

11.1 Ballistic pendulum assembly

1. With scissors, make an opening measuring approximately 9 inches x 9 inches in the front of the box, as shown in the box detail diagram. Reinforce all ends with duct tape. Fill the box with wadded-up newspapers and weights to make it weigh exactly 64 ounces (4 pounds).

2. Attach a string to each upper corner of the box by attaching small eye hooks to the corners. Reinforce the eye hooks securely with duct tape.

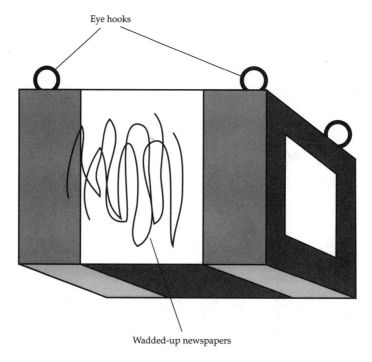

Eye hooks

Wadded-up newspapers

11.2 Ballistic pendulum box detail

3. Attach the felt tip marker to the box so it protrudes just beyond the side plane of the box. Refer to the assembly diagram.

4. Attach the sheet of paper to the plywood and mount the pendulum and plywood parallel to the pendulum's path. Position the paper so it just touches the felt tip marker.

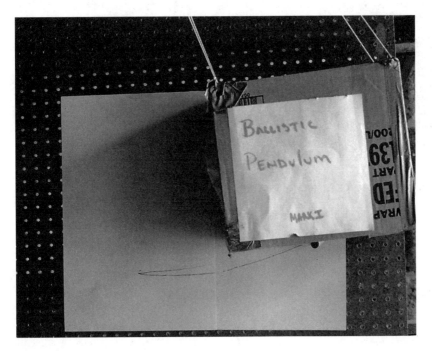

11.3 Box with felt tip marker attached, paper photo

SETTING UP THE PENDULUM

Refer to the ballistic pendulum assembly diagram. Suspend the pendulum from a ledge or ladder as shown. The general idea to keep in mind is that the cannon (or mortar, catapult, etc.) will be shot point-blank into the ballistic pendulum. The pendulum will swing after the projectile collides with it. When the projectile is fired, its momentum is transferred to the pendulum and its velocity can be determined from the height to which the pendulum rises.

It is important that the box swings freely after colliding with the projectile. As it swings, the marker will trace out the path of the box on the paper that is attached to the plywood board, marking the pendulum's highest and lowest points. The difference in height between the apex of the swing and the rest position allows us to determine the speed of the projectile.

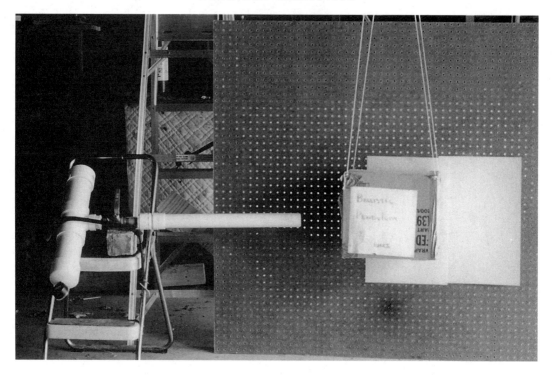

11.4 Ballistic pendulum

DETERMINING MUZZLE VELOCITY

1. Refer to the assembly diagram and photo. Mount the
   potato cannon directly in front of the cardboard box so
   that the projectile will be fired into the middle of the
   opening of the box and into the wadded-up newspapers.
   Make sure the pendulum swings in a balanced fashion
   and that the marker will trace out a clean arc on the
   paper. Uncap the felt tip marker.
2. Carefully weigh the potato projectile on a postal scale,
   and record the amount to the nearest half-ounce.
3. Fire the cannon into the pendulum.
4. Now, measure the difference in height between the highest point of the arc traced by the felt tip marker on the
   paper, between the apex and the rest position. See the diagram to understand how to measure the height correctly.

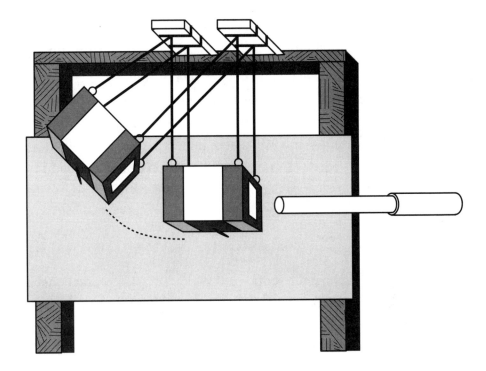

11.5 Ballistic pendulum after firing

5. With the three parameters now known—the mass of the projectile, the mass of the pendulum, and the height to which the pendulum rises—you can accurately calculate the muzzle velocity of the projectile.

PHYSICS

Earlier in this book, Sir Isaac Newton's three laws of motion were discussed. Let's apply those laws of motion to determine the velocity of a bullet.

Physicists would say that the collision between the spud and the pendulum is "perfectly inelastic" because the wadded-up newspapers allow for no bounce or rebounding whatsoever when the potato hits the pendulum. In this type of collision, physicists would say that "momentum is conserved." However, in this type of collision, where there is no bounce or elasticity between the colliding objects, those same scientists would note that the kinetic energy is not conserved. Energy and mass conservation laws are the basis of analysis that scientists use to relate mass, velocity, distance, and time. Through careful choice of the variables involved, and mathematical manipulation of the physics equations that describe the process, an equation can be written to let us determine the speed of the projectile from the easily measured variables discussed earlier.

$$V \text{ potato} = \left(1 + \frac{M}{m}\right) \sqrt{2gh}$$

It is beyond the scope of this book to derive the equations for muzzle velocity. Most beginning physics textbooks discuss the physics of the ballistic pendulum when one-dimensional particle kinetics and energy and momentum conservation laws are introduced. For now, take it on faith that the muzzle velocity of the potato is given by this equation:

Where M is the weight of the pendulum, m is the weight of the potato, g is the earth's gravitational constant, and h is the vertical rise of the pendulum.

If we insert the constants, we can make a fairly easy-to-use equation to determine the speed of any projectile shot into our pendulum:

$$V = \left(1 + \frac{M}{m}\right) \sqrt{43.9 \times h}$$

▶ The velocity of projectile is in miles per hour.
▶ The weight of projectile and pendulum can be ounces, pounds, or grams, but units must be used consistently.
▶ The height of pendulum rise is in feet (or use inches and divide by 12).

To find the muzzle velocity of any projectile, substitute your figures in the equation given.

TIPS AND TROUBLESHOOTING

1. A weight of four pounds works well for the pendulum when measuring muzzle velocity of a potato cannon. You can use lead fishing weights in a small bag or something similar to make the box weigh as close to four pounds (64 ounces) as possible.

2. In order to get an accurate reading of the muzzle velocity, it is very important that the pendulum and potato be measured very accurately. Use a postal scale to determine weights to a half-ounce or better.

3. Take care to line up the marker exactly perpendicular to the paper. The marker tip should rest squarely and lightly on the paper.

▶▶ IN THE SPOTLIGHT ◀◀

BENJAMIN ROBINS, INVENTOR OF THE BALLISTIC PENDULUM

Benjamin Robins was a hardworking and well-traveled engineer whose body of work provides some of the most important scientific foundations for the study of ballistics. He grew up in Bath, England, and studied sciences and engineering in London. Somewhat of a prodigy, he was published in the *Philosophical Transactions of the Royal Society* in 1727, and was elected a Fellow of the Royal Society when he was only in his mid-20s.

Robins left the academic life to become an engineer. His construction projects included civil engineering projects of all kinds:

factories, bridges, mills, and so forth. In addition, he began to study military science. He traveled through Europe to gain experience, walking through the great castles and forts of France and southern Europe, taking notes and making drawings.

On his return to England, he published *A Discourse Concerning the Nature and Certainty of Sir Isaac Newton's Method of Fluxions*. Robins's book solidified Netwon's reputation. Newton, as you'll recall, was a great but controversial scientist. His contemporaries seemed to side strongly with him (like Benjamin Robins) or against him (like Robert Hooke). Robins was one of Newton's most stalwart supporters, and his scholarly works boosted Newton's claims and legitimacy.

In 1742, Robins's *New Principles of Gunnery*, arguably his most important work, was published. This landmark text formed the basis for all subsequent work on the theory of artillery and projectiles. For this work, he received a very high honor, the Copley Medal of the Royal Society. In *New Principles of Gunnery*, Robins built on the work of an Italian, J. D. Cassellini, who researched and published on the subject about 30 years earlier. In *New Principles*, Robins first develops and explains the ballistic pendulum. This device allowed precise measurements of the velocity of projectiles fired from guns. Just like our ballistic pendulum, Robins suspended a large wooden block in front of a gun and measured the height it attained after colliding with a projectile.

Robins was the consummate military engineer. A man of many talents, he experimented with rockets, publishing *Rockets and the Heights to Which They Ascend* in 1750. His experience and skill made him a valuable military advisor, and the king called him to service once more for a mission to Madras, India, to improve British colonial defenses in 1750. Unfortunately, the Indian climate was not good for Robins. There, he contracted a fever and died.

Ideas for Further Study

IDEAS FOR FURTHER STUDY

If you want to further explore the physics behind these experiments, here are some ideas to pursue. You'll need to incorporate the *scientific method* into your plans so you can accurately understand and communicate the significance of your discoveries. In general terms, in order to prove a cause and effect relationship, you must test a single variable and condition at a time.

The Scientific Method

On occasion, a discovery is made thorough a lucky accident (think of the Chinese and the way they discovered the effect of saltpeter in gunpowder), but more often the scientist must carefully develop a procedure to test the validity of a theory or idea. The way scientists go about proving or disproving an idea is called the "scientific method." There are four major parts to the scientific method that are always incorporated into a scientific investigation—hypothesis, procedure, data collection, and conclusion.

HYPOTHESIS

Choose a topic you would like to explore. Often the problem starts off with a question, such as, "At what angle should I tilt the potato cannon in order to make the potato fly the farthest?" Once you have a question you would like to explore, you need to restate it in the form of a hypothesis.

A hypothesis is a statement that describes how a variable will affect an event. To make the above question into a hypothesis, you could restate it this way: "In order to achieve the greatest distance, I will place the cannon at a 45-degree angle from the ground." It's important that your hypothesis be very clear so that you can test it. In this example, what you propose to test is that a 45-degree angle results in the greatest distance.

PROCEDURE

This section should describe what you plan to do during your experiment. List all the materials you will need. Then list each task you will need to do, in order. Number each task. Write down everything you will do. Other scientists should be able to repeat your experiment by reading your procedure. For the potato cannon example, your materials would include potatoes, the cannon, a protractor, a marker for recording distances, and a clipboard.

DATA COLLECTION

As you perform the steps exactly as described in the procedure, you should write down your observations. These are the data. Record your observations as accurately as possible. You may want to organize your data into a table format to make it easier to record and analyze. In the potato cannon project, for example, try several potato firings at angles of, say, 35, 45, 55, and 65 degrees. After firing several spuds at each angle, take note of the distance covered at each angle.

CONCLUSION

Look carefully at your data and decide what it tells you about your hypothesis. For example, you can graph the results of each angle, plotting the firing angle on the x-axis and the distance covered on the y-axis. The highest point of this graph tells you which trajectory yields the greatest distance.

You may decide at this point that you need to revise your hypothesis and think about further experiments. You may also decide to communicate your results to others in a scientific article, which is how scientists let others know of their work.

Playing with Fire—Los Alamos Style

Scientists use the scientific method for very large and involved experiments, too. Consider the Cold War–era experiment that resulted in the first nuclear-powered spud gun.

In 1955, a group of scientists and engineers at Los Alamos National Laboratory were given the task of reducing the amount of radioactive material expelled into the atmosphere resulting from nuclear testing. Astrophysicist Robert Brownlee was a principal participant in these tests, named project Bernalillo after a New Mexico county near Los Alamos.

Dr. Brownlee and his team were testing the feasibility of moving nuclear testing underground. In order to achieve a number of scientific objectives, they needed to explode several nuclear devices underground. Doing so involved building the equivalent of a giant, atomic-powered potato cannon. The cannon was a 400-foot-deep well lined with thick steel pipe, capped by a steel plate instead of a potato, and pow-ered by a nuclear bomb instead of a squirt of hairspray.

Forty stories below the scrubby tangle of mesquite trees and creosote on the desert surface, researches tried to deter-mine if they could safely test the effects and design of nuclear devices while reducing the release of radioactive materials into the atmosphere to a minimum, maybe even completely.

The Bernalillo team placed a small (by high-energy physics standards) nuclear device in the steel well and capped the well off with a big steel manhole cover. The four-foot-diame-ter steel manhole cover was four inches thick and weighed in the neighborhood of half a ton.

This puny nuclear device had the explosive equivalent of less than one kiloton of high explosive. However, small in nuclear terms is still incredibly large. The effects of letting lots and lots of nuclear energy loose are sometimes hard to predict. To understand what happened when the device was triggered, the Los Alamos team utilized the scientific method. They started with a hypothesis and then assembled an array of state-of-the-art measuring equipment to test it.

The scientists working on the Bernalillo series of test shots were trying to figure out what happens during the few micro-moments of the nuclear explosion. The Los Alamos team wanted to know what kind of nuclear particles were emitted, how many there were, and, most importantly, where they went. The data they needed to collect had to be meas-ured in the first few *shakes* after the explosion begins. (A "shake" is the amount of time it takes light to travel 10 feet. Since light travels at around 186,000 miles per second, that makes a shake an exceedingly short time interval.)

The scientists put all sorts of detectors and sensors in and near the well. They also placed high-speed cameras some dis-tance from the top of the well to film the explosion. Normal cameras take about 16 frames of film every second. The high-speed Los Alamos cameras were 10 times faster.

When the device was triggered, the scientists got a bit more than they bargained for. The bomb emitted high-energy particles of light, called photons. Within a few shakes, the photons, or in Alamos lingo, the "shine," bombarded the steel pipe, vaporizing it into superheated iron gas. About three hundredths of a second after detonation, the shock wave of gas, light, and radiation blasted against the steel cover plate at the top of the well.

The high-speed cameras recorded the blast effect on the plate. In one frame the plate is there. In the very next frame, $1/160^{th}$ of a second later, it is gone. Where did the four-foot diameter, heifer-sized steel plate go? The area was searched carefully, but the plate wasn't found. In fact, in the 40-plus years since project Bernalillo, no trace of the plate has ever been found, anywhere.

The project team felt they knew where the plate went. Prior to the actual test, Dr. Brownlee's boss asked him what would happen to the plate covering the test hole. He thought about it for a while. "I guess I don't really know," said Brownlee. "Find out," said the project director.

Brownlee performed some preliminary calculations. Based on the expected bomb yield, the shape and depth of the test hole, and so forth, he figured the initial velocity of the plate would be somewhere in the neighborhood of 41 miles per second. That's moving mighty fast. He made many slide-rule calculations and reported back to his director. The manhole cover would probably wind up on a collision course with the distant stars, shoved by a nuclear push through the Earth's atmosphere and into outer space.

In 1687, Isaac Newton figured out some interesting things about gravity and velocity. He deduced that there is one particular speed, one where if you throw something hard enough and fast enough, you can make it through the gravitational attraction of the Earth and break free into outer space. Newton called this speed "escape velocity," and on Earth this is calculated to be just a hair less than seven miles per second. When the Bernalillo team calculated the plate's velocity just after detonation, they estimated it was in the rough neighborhood of *five times* escape velocity! Even taking into account possible assumptive errors and other unknowns, it seemed likely that the plate was traveling well above the speed required to escape the gravitational force of Earth.

To test the validity of Brownlee's calculation, other Los Alamos scientists reviewed the film from the high-speed cameras. Upon review, they found the plate was present in one frame of the high-speed film and gone in the next. They factored in what they new about the film speed and the field of view of the camera. Based on the photographic evidence, the scientists felt a strong case could be made that the half-ton steel plate was moving faster—in fact, much faster—than escape velocity.

A few years later, in 1959, a team of Soviet scientists launched what they claimed to be the first man-made object into outer space, the satellite *Sputnik*. Many people at Los Alamos think *Sputnik* was merely the second object to travel to outer space, preceded by a full two years by an American-made manhole cover.

The Bernalillo project illustrates all four parts of the scientific method. The Los Alamos scientists began with a hypothesis. In rough terms, it went something like this: Testing nuclear devices underground will reduce radioactive emissions into the atmosphere.

Next, they came up with a procedure. This step included the work of designing the well, the bomb, the cover plate, and so on. And it included figuring out how, when, and where instrumentation would be used to collect the required data. Sensors, cameras, and particle detectors were designed and placed to measure the types and amount of radioactive particles, blast pressures, and temperatures, and so on.

After the data was collected, it was analyzed, and the scientists were able to draw their conclusions. Yes, underground testing was a viable way to test these devices and reduce airborne radioactivity at the same time.

QUESTIONS TO EXPLORE

CINCINNATI FIRE KITE

1. What is the effect of humidity and temperature on the ascent rate of a Cincinnati fire kite? Graph the rate of ascent under different weather conditions. Did the kite rise faster when the weather was cool or warm? What effect did high relative humidity have?
2. Use a newspaper sheet of different sizes. What effect does size have on the ultimate height attained by the kite?

CATAPULT

1. What is the relationship between the number of twists in the torsion bundle (the twisted spring) engine of the catapult and the distance a projectile is thrown?
2. Make a graph of the number of twists in the bundle and the distance. Do the results plot in a straight line? If it does, then there is a linear relationship between the amount of twist and the distance a weight is tossed.

TENNIS BALL MORTAR

1. Make a tennis ball mortar without the middle baffle. What effect does baffling have on distance?
2. Try using baffles of different diameter. How well does the mortar work with approximately 40 percent of the area cut away? What about 60 percent? 80 percent? What is the optimum area of baffling for best performance?

POTATO CANNON

1. Try ramming the potato about 12, 24, or 36 inches down the muzzle. At which depth does the cannon shoot the potato the farthest distance?
2. Place the potato cannon at an angle of 25 degrees from horizontal. Spray a carefully measured two-second charge into the cannon and fire. Measure the horizontal distance traveled. Repeat the firing at angles of 45-degree and 75-degree from the horizontal. Which angle of fire provides the greatest distance?

MATCH ROCKET

1. Trim the matchstick paper into different shapes and lengths. Which design gives the best performance?
2. Make a double match rocket by wrapping two match heads together. Does the rocket fly twice as far? Half as far? Why?
3. Use a wooden kitchen match instead of a paper match in one of your rockets. How well does it fly?

Index

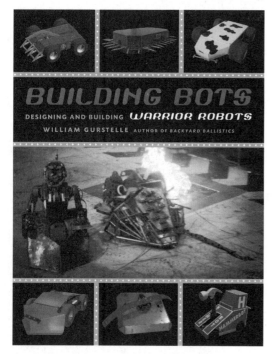